WHAT EVERY ENGINEER SHOULD KNOW ABOUT
Lasers

WHAT EVERY ENGINEER SHOULD KNOW
A Series

Editor

William H. Middendorf

*Department of Electrical and Computer Engineering
University of Cincinnati
Cincinnati, Ohio*

Vol. 1 What Every Engineer Should Know About Patents, *William G. Konold, Bruce Tittel, Donald F. Frei, and David S. Stallard*

Vol. 2 What Every Engineer Should Know About Product Liability, *James F. Thorpe and William H. Middendorf*

Vol. 3 What Every Engineer Should Know About Microcomputers: Hardware/Software Design: A Step-by-Step Example, *William S. Bennett and Carl F. Evert, Jr.*

Vol. 4 What Every Engineer Should Know About Economic Decision Analysis, *Dean S. Shupe*

Vol. 5 What Every Engineer Should Know About Human Resources Management, *Desmond D. Martin and Richard L. Shell*

Vol. 6 What Every Engineer Should Know About Manufacturing Cost Estimating, *Eric M. Malstrom*

Vol. 7 What Every Engineer Should Know About Inventing, *William H. Middendorf*

Vol. 8 What Every Engineer Should Know About Technology Transfer and Innovation, *Louis N. Mogavero and Robert S. Shane*

Vol. 9 What Every Engineer Should Know About Project Management, *Arnold M. Ruskin and W. Eugene Estes*

Vol. 10 What Every Engineer Should Know About Computer-Aided Design and Computer-Aided Manufacturing: The CAD/CAM Revolution, *John K. Krouse*

Vol. 11 What Every Engineer Should Know About Robots, *Maurice I. Zeldman*

Vol. 12 What Every Engineer Should Know About Microcomputer Systems Design and Debugging, *Bill Wray and Bill Crawford*

Vol. 13 What Every Engineer Should Know About Engineering Information Resources, *Margaret T. Schenk and James K. Webster*

Vol. 14 What Every Engineer Should Know About Microcomputer Program Design, *Keith R. Wehmeyer*

Vol. 15 What Every Engineer Should Know About Computer Modeling and Simulation, *Don M. Ingels*

Vol. 16 What Every Engineer Should Know About Engineering Workstations, *Justin E. Harlow III*

Vol. 17 What Every Engineer Should Know About Practical CAD/CAM Applications, *John Stark*

Vol. 18 What Every Engineer Should Know About Threaded Fasteners: Materials and Design, *Alexander Blake*

Vol. 19 What Every Engineer Should Know About Data Communications, *Carl Stephen Clifton*

Vol. 20 What Every Engineer Should Know About Material and Component Failure, Failure Analysis, and Litigation, *Lawrence E. Murr*

Vol. 21 What Every Engineer Should Know About Corrosion, *Philip Schweitzer*

Vol. 22 What Every Engineer Should Know About Lasers, *D. C. Winburn*

Other volumes in preparation

WHAT EVERY ENGINEER SHOULD KNOW ABOUT
Lasers

D. C. WINBURN

Los Alamos, New Mexico

MARCEL DEKKER, INC. New York and Basel

Library of Congress Cataloging-in-Publication Data

Winburn, D. C.
  What every engineer should know about lasers.

  Includes index.
  1. Lasers. I. Title.
  TA1675.W56   1987   621.36'6   87-5279
  ISBN 0-8247-7748-4

COPYRIGHT © 1987 by MARCEL DEKKER, INC. All Rights Reserved

Neither this book nor any part may be reproduced or transmitted in any form or by any means, electronic or mechanical, including photocopying, microfilming, and recording, or by any information storage and retrieval system, without permission in writing from the publisher.

MARCEL DEKKER, INC.
270 Madison Avenue, New York, New York 10016

Current printing (last digit):
10 9 8 7 6 5 4 3 2 1

PRINTED IN THE UNITED STATES OF AMERICA

*To Vicki, Barry, and Randy, my children
whose lives will be touched in many ways by lasers*

Preface

The laser has been called the "supertool of the 1980s" by Hecht and Teresi,* the "light fantastic" by some, and other endearing terms by many other people in describing the fabulous feats of this "high-technology" invention, created by the early masters of science, particularly Albert Einstein. Laser devices have only recently (in the past two decades) been developed to be of practical use in countless ways to benefit mankind. It is in many ways an invention waiting to be utilized—in science, industry, medicine, communications, entertainment, and, some believe, as a deterrent to military aggression by threatening instant retaliation, as in the "star wars" concept of advanced laser technology.

The versatility of the "photon," the fundamental ingredient of laser light, extends from the ultra-low energies required in biology research to study individual atoms of the DNA molecule and separate cancer cells from healthy cells to the creation of temperatures found only in stars by focusing huge amounts of laser energy onto small thermonuclear targets, and, even further, to the physical destruction of specified targets. So many applications are evolving that technical books, periodicals, and other publications cannot disseminate the latest information quickly enough to the many disciplines of technology of our modern society. Details of research and applications development must be obtained at conferences or meetings of special interest groups such as the International Society of Photo-

---

*Jeff Hecht and Dick Teresi, *Laser, Supertool of the 1980s* (1982). Ticknor & Fields, New Haven, CT.

optical Instrumentation Engineers (SPIE), the Optical Society of American (OSA), the Laser Institute of America (LIA), the Institute of Electrical and Electronics Engineers (IEEE), small surgical and medical societies, and metals and materials specialists.

The basic physics of the laser has been published in several forms, including textbooks for college and vocational study, but there is a need for those practical "doers," the engineers, to have a single source to explain in simple, straightforward terms the fundamental workings of lasers and how coherent light interacts with various materials, and to describe in concise terms many of the myriad applications currently in use. That is the purpose of this book—to present current laser technology and applications in a manner understandable by engineers of various disciplines. As a metallurgical engineer with 40 years of experience, having been active in the research and development of nuclear and thermonuclear materials, and having been a technical administrator in laser research and technology, it is my belief that basic laser technology can be communicated in an acceptable form to fellow engineers—at least the information all engineers should know to understand the potential of this fabulous tool in their particular field.

D. C. Winburn

Contents

Preface	v
1. Introduction	1
2. Types of Lasers	5
Solid-State Lasers	7
Semiconductor Lasers	9
Gas Lasers	11
Chemical Lasers	13
Liquid Lasers	15
Excimer Lasers	17
Free Electron Lasers	18
Other Laser Types	19
References	20
3. Characteristics of Laser Beams	21
Electromagnetic Spectrum	21
Wavelength Effects	33
Continuous Beam (CW) Properties	34
Pulsed Beam Traits	35

	Beam Cross-Sections, or "Modes"	38
	Beam Divergence	42
	Beam Focusing	43
	References	47
4.	**Measurement of Laser Beam Characteristics**	**49**
	Wavelength Measuring Devices	49
	Power and Energy Measurements	62
	Photographic Instrumentation	76
	Bibliography	82
5.	**Safety in the Laser Environment**	**83**
	Laser Beam Characteristics	84
	Biological Damage Thresholds	87
	Engineering Controls	88
	Indoctrination and Training	91
	Administering a Laser Safety Program	93
	Summary of Practical Laser Safety Program	95
	References	96
6.	**Engineering and Science Applications**	**97**
	Materials Processing	98
	Communications Using Fiber Optics	109
	Holography	115
	Chemistry	118
	Construction Industry	124
	Computers and Printers	124
	Micro-Wire Stripping	128
	Astronomers' Measurements	129
	Doppler Effect Measurement	130
	Laser Tunnelling	131
	Laser Alignment System	131
	Laser Detector for Search and Rescue	132
	Shock Wave Diagnostics	133
	Guidance by Laser Gyro	133
	Laser Wind-Sensor	134
	Fingerprint Identification	135
	Sound Recovery from Antique Audio Cylinders	136
	Laser Disc Recorders and Players	138
	Video Discs	139
	References	140

CONTENTS

7.	**Medical Applications**	141
	Surgery	143
	Ophthalmology	150
	Dermatology	155
	Cell Sorting	156
	Other Medical Applications	158
8.	**Laser as an Art Form**	165
	Government Requirements	167
	Open Air Laser Light Shows and FAA Requirements	169
	State and Local Laser Light Show Requirements	170
	Individual Responsibilities	174
	What the Public Should Know about Laser Safety	176
9.	**Future Developments of Laser Technology**	177
	Automobile Navigation by Lasers	177
	Military Applications	179
	Laser-Induced Fusion	190
	Nuclear Laser	195
10.	**Summary**	197
Index		199

WHAT EVERY ENGINEER SHOULD KNOW ABOUT
Lasers

1
Introduction

Any system that can be applied to such diverse technologies as welding metal and eye surgery is bound to do quite well in the marketplace. So it is with lasers, which have finally shed their "science fiction" aura and are finding their way into nearly every sector of the world economy.

Although laser technology has been with us since the 1960s, the full potential of this magnificent tool has yet to be realized. The domestic market for all types of nonmilitary lasers stood at $210 million in 1982—an increase of nearly tenfold from 1972—but this is only the beginning, according to Predicasts, Inc., the Cleveland-based market research and business information firm.

In their recently published report, *U.S. Laser System Markets*, Predicasts projects U.S. nonmilitary laser systems sales to reach $665 million by 1987 and a whopping $1.6 billion by the mid-1990s. More impressively, says Predicasts' Research Analyst, Edward Hester, laser sales will grow at a pace roughly three times that of the nation's gross national product over the same period.

Until recently, the fascinating intricacies of laser technology was limited to the military, academia, and laboratory researchers. Now, however, the commercial exploitation of laser technology across numerous areas of industry, communications, science, and medicine is gaining momentum.

With $80 million in sales, communications comprised the largest market for lasers in 1982, specifically diode lasers used in fiberoptic communication and helium neon lasers as applied in reprographics, printing, and point-of-sales (POS) bar code scanners. Laser sales to the communications segment are expected to maintain a 28 percent annual increase in the near term, then level off in the late 1980s as unit price declines take hold. Still, the growing demand for high-tech communication systems will boost the market to over a half billion dollars by 1995.

The recession of 1981-1982 devastated many manufacturing-oriented industries, but the slump's effect on industrial lasers was to hold growth in this segment to "only" 17 percent per year from 1972 to 1982. The $60 million spent for industrial lasers in 1982 covered a number of applications, including welding, drilling, cutting, heat-treating annealing, and lithography. Laser manufacturers and distributors are expected to benefit particularly well from the ongoing economic recovery, based on the ability of lasers to provide increased productivity, pinpoint precision, and reduced labor and energy costs. By 1987, Predicasts projects the industrial market for lasers to reach $165 million, a pause on the way to the $400 million mark by the middle of the next decade.

Medicine and science are expected to be the fastest-growing end-use markets over the next decade, particularly since lasers can be used to perform bloodless surgery and are expected to play an increasingly greater role in the fight against cancer. The medical and scientific markets bought $70 million worth of laser equipment in 1982—up nearly 26 percent annually since 1972. Predicasts sees this market growing at 31 percent per year through 1987, resulting in $275 million in sales. By 1995, medical and scientific lasers are expected to account for 40% of the total laser market, or $655 million.

Healthy growth is anticipated through the study period for all laser types, including solid state, gas, and dye. Solid-state types will command the lion's share in the short term—32 percent per year through 1987—as the nation's communications systems continue to convert from narrowband analog to broadband digital. For the remainder of this century, copper wires will be supplanted by optical fibers, providing excellent opportunities for laser diodes and related equipment. By 1987, solid-state lasers will hit the $360 million mark, heading toward a market of $750 million by the mid-1990s.

Gas lasers achieved sales of $110 million in 1982, up from $16 million in 1972. Increased medical application of gas lasers over this period helped overcome some softness in the industrial market as the economy slumped in the early 1980s. Because they are so prevalent in materials processing, which traditionally is one of the last sectors to rebound during an economic recovery, the growth of gas lasers is expected to trail that of other laser types for the next three to five years. However, since they are flexible enough to operate at almost any power level and have good tunability, gas lasers offer the best long-term prospects. Predicasts expects gas lasers to represent a $275 million market by 1987, and $750 million by the mid-1990s.

Dye lasers, whose primary applications are in test and measurement, will remain the smallest segment of the laser industry through the study period, says

INTRODUCTION

Table 1.1 U.S. Laser System Markets

					% Annual Growth		
Item	1972	1982	1987	1995	72-82	82-87	87-95
GNP (bil 72$)	1085.9	1476.9	1790.0	2330	3.1	3.9	3.4
GNP deflator (1972 = 100)	100	207.1	270.8	415.0	—	—	—
GNP (bil$)	1085.9	3059.3	4847.0	9670.0	10.9	9.6	9.0
laser sales/mil$ GNP	23.9	68.6	137.2	163.4	—	—	—
New P&E expend (bil$)	120.3	316.4	500.0	1045.0	10.2	9.6	9.7
laser sales/000$ P&E	21.6	66.4	133.0	151.2	—	—	—
Laser sales (mil$)	*26*	*210*	*665*	*1580*	23.2	25.9	11.4
By type:							
Solid state	10	90	360	750	24.6	32.0	9.0
Gas	16	110	275	750	21.3	20.1	13.4
Dye	neg	10	30	80	—	24.6	13.0
By end-use:							
Industrial	12	60	165	400	17.5	22.4	11.7
Communication	7	80	225	525	27.6	23.0	11.2
Medical & scientific	7	70	275	655	25.9	31.5	11.5

Source: Predicasts, Inc., 11001 Cedar Avenue, Cleveland, Ohio 44106.

Predicasts. The market for dye lasers is expected to reach $30 million by 1987—up from $10 million in 1982. Growth for dye lasers will average 13 percent per year between 1987 and 1995, when sales of $80 million will be realized.

The U.S. laser systems markets listed in Table 1.1 do not include the military market, which is expected to grow at a rapid rate to reach the success anticipated in the "Star War" concept. Military use of laser technology could match domestic use in the near future.

These predictions reveal the recent upsurge of intense interest in lasers in essentially all technical fields. If an engineer is to keep abreast of current and evolving laser applications, somewhat more than a cursory knowledge of what lasers are and how the photon beam interacts with materials must be accumulated to keep competitive in the technical field of interest. By learning the fundamental characteristics of the several types of lasers and the accessory technologies assigned to specific types, adapting a photon beam to a particular process or function can be evaluated. The knowledge required by some engineering applications of laser systems can be quite extensive, and that sort of in-depth information is available from references contained in this work; but, for the "average" engineer not having any hands-on, direct experience with lasers, a more limited exposure to the technical aspects of this emerging technology would seem adequate—and certainly necessary.

Although certain technical aspects of lasers will be presented, especially in the beam characteristics and beam measurements sections, the major emphasis will be on applications of lasers requiring only limited mathematical and scientific acumen to understand the applications.

2
Types of Lasers

Included here is a brief summary of how the understanding of light and its properties progressed over the centuries.

According to the *Encyclopaedia Britannica*'s historical survey of light, from 500 B.C. to A.D. 1650, there were innumerable confusions and false starts toward an understanding of light. From 1650 to 1895, several renowned physicists unravelled many fundamental properties of light, including the existence of a connection between electromagnetism and light proposed by Maxwell and demonstrated by Faraday and others. In fact, late in this period, the value of c, the speed of light in vacuum, was determined from measurements on electrical circuits as 300,000,000 (3×10^8) meters per second by A.A. Michelson, a U.S. physicist. From 1900 to the present, the theory of light has reached a point at which all terrestrial phenomena are included in one logical theory: Maxwell's theory of waves in a continuous medium. Planck demonstrated that it is necessary to postulate that radiant heat energy is emitted only in finite amounts, which are now called "quanta." In 1905, Einstein showed that in the photoelectric effect, light behaves as if all the energy were concentrated in quanta, particles of energy now called photons. In that same year, Einstein published the theory of relativity, which modified the whole of physics and gave a special role to the light velocity constant, c, in the famous formula $E=mc^2$. In 1916 [by some authorities (1)], Einstein conceived a theory that if an electron were in an excited state when a photon previously emitted and having the proper energy collided, the electron would drop to a lower energy state, and emit another photon of the same energy that would move in the same direction, resulting in two identical photons travelling together in the same direction and same phase. The term *stimulated emission* was established by this theory. This theory, and previous good works, were the basis of the grant of the Nobel prize in physics to Einstein in 1921. It is also the fundamental explanation of how energy is generated in a coherent, single-wavelength, narrow beam of light known by the acronym LASER (*L*ight *A*mplification by *S*timulated *E*mission of *R*adiation). This chapter will discuss briefly how emission from the various types of lasers is generated and provide an explanation of how some of the fundamental properties of the various materials used in lasers contribute to the wavelength, or color, of the beam.

Einstein's theory, with a contribution from an English physicist, Dirac, postulated that under certain conditions atoms could be made to radiate in phase so that highly coherent (same direction and same wavelength) radiation could be maintained indefinitely. The practical realization of these conditions, previously thought to be impossible, was achieved in 1960. The history of development of the first working lasers has been recorded in detail in other works, and will not be attempted here, but the unusual progression of laser technology makes for interesting reading about the independent, simultaneous research accomplished by American and Soviet physicists and about the conflicting claims for patent rights by several inventors.

Laser technology is developing at such a pace that discoveries of new lasing materials, advanced lasing methods, and new laser applications are reported by the news media almost every week. All the states of matter can produce laser radiation, but the most common in current vogue are the solid-state lasers followed closely by gas lasers and then, to a lesser extent, the liquid laser media. Often laser researchers report newly discovered materials and techniques that have no immediate application, but before long evaluation of the advances will give the "new beam" a place in our current technical world. Because each type of laser beam is unique in its effect on its absorption by or transmission through various materials, the beam must be thoroughly investigated to determine its desirability of use in various applications. For example, the ruby laser beam (first discovered by Maiman in 1960) is transmitted through many optically clear materials such as glass and plastics but is absorbed by any opaque or dark substance, so that, in one application at Los Alamos, the ruby laser beam is used to penetrate a glass vacuum system without harm, yet when focused on a small thin-walled metal sample container it can melt a hole in the container to release gas species produced within the container yet maintain the integrity of the vacuum system. This particular experiment included passage of the laser beam through a foot-thick leaded window of a "hot cell" containing radioactive materials before entering the vacuum system. This application illustrates the usefulness of a laser of certain specific wavelength with adequate power density and having the desired properties

SOLID-STATE LASERS

of absorption and transmission. Also involved in the selection of lasers are the services required to operate the laser, the life of active use expected, and cost effectiveness. Although these factors will be discussed in this chapter, it should be noted that those systems just entering the market with relatively short experience records must be carefully evaluated before adapting such systems to the desired use. The reputation and history of the manufacturer should be known, and written long-term warranties are recommended. The changing nature of the laser market has a history of volatility and the legal term "caveat emptor" (let the buyer beware) particularly applies to this industry.

The type of laser selected for a particular application may be the result of experimenting with various wavelengths to determine adaptability, particularly if laser radiation is to be an integral component of a permanent installation. Consequently, the type of lasing medium may be dictated by the experimentation because the laser technology has not advanced to the point where the laser user can specify the preferred type. For example, the helium-neon laser, a gas type, has many alignment applications but is limited at the present time in power output. Because of this limitation, a krypton (gas-type) laser at a slightly longer wavelength was employed at Los Alamos to provide a higher power density potential in alignment of high energy CO_2 laser radiation that traversed through gas chambers and other absorbing gear in its path to the target chamber. Most reputable laser manufacturers will help potential customers who are inexperienced in laser technology by providing consulting services to ensure optimum efficiency in the application at hand. However, an independent reference (2) for the safe use of lasers is recommended before use of any laser.

Of the various types of lasers currently in use for a plethora of applications, the following are the most important.

SOLID-STATE LASERS

The materials involved in producing a cascade of photons from this type are, as the name implies, solids. The actual photon-emitting atoms are generally from a small percentage of an element in a matrix of other compounds. For example, the ruby laser obtains its photons from the element chromium, which is only a few percent of the aluminum oxide matrix. The YAG laser radiation results from lasing of the neodymium atoms which are a few percent of the ytrium-aluminum-garnet matrix. Although the natural gem materials can be used for these two solid-state lasers, the lasing media have been synthesized and produced on a large scale for high-rate laser manufacturing and improved material quality.

How do solid-state lasers work? The diagrammatic sketch in Figure 2.1 makes it look simple to generate the ruby laser beam, but this was not the initial design anticipated by the early investigators who believed the first lasers would

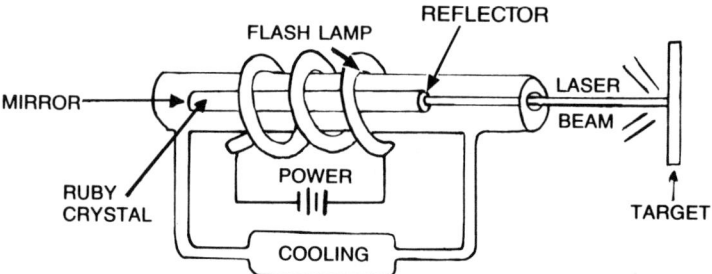

Figure 2.1 Diagrammatic sketch of a ruby (solid-state) laser. Essentially all solid and gas lasers have these common components: (a) laser cavity solids or gas chamber; (b) mirrors at ends (one of which is a partial transmitter); and (c) activation mechanism—either a light source or electric discharge facility.

be made by passing an electric current through a gaseous media to obtain the required excitation. However, laser pioneer Theodore Maiman, working alone at the Hughes Research Lab in California, discovered in 1960 that the chromium atoms could be excited by a flash lamp. This process is now described by the expression "optically pumping," because energy is pumped into the atomic structure of the laser material for excitation to a higher energy level in order for laser action to take place. The illumination of the chromium atoms occurs with light of a frequency* higher than that which the laser is to omit. This light pump must be of high intensity because the process itself is inefficient.

Referring to the diagram for the optically pumped ruby laser (Fig. 2.1), notice that the ends of the rod are polished flat and parallel, then coated with reflecting material. The sides are left clear to admit the light for excitation. The lamp may be pulsed and must be cooled for temperature control. The light source may be wound around the rod or placed alongside, or even focused on it by mirrors. The laser beam may be "switched" by a device called a "Q-switch control" for obtaining high output pulses of laser light at very short pulses, ~ one nanosecond (10^{-9} sec). The beam is propagated by reflection of those photons generated by stimulated emission travelling normally to the mirrored ends. As the intensity builds owing to the action of the speed of light, the "open" end of the rod, which is only partially mirrored, releases the photons in a stream, the photons all moving in the same direction and exactly in phase (coherent) with high intensity. Incidentally, the distance between the mirrored ends of the rod must be precisely positioned to reflect properly at exact wavelength distances to preclude interference from reflected photons. An important factor in the design of

---

*Frequency is the inverse of wavelength, so that if a frequency is doubled, the wavelength is halved.

SEMICONDUCTOR LASERS

this laser type is a cooling system to keep the lasing material from deterioration, because only a small fraction of the light energy is transferred to laser energy. The volume occupied by the laser material is referred to as the "laser cavity." In solid lasers, the cavity is the volume of the solid.

The first operating ruby lasers used artificially produced sapphire crystals (aluminum oxide). Many other rare earth elements have been used in this process, containing small percentages ($\sim$2-3%) of laser materials in various matrices. The most popular laser of the optically pumped solid type is the neodymium (Nd) "doped" glass, referred to as the YAG laser because the original natural matrix, or "host" material, consisted of yttrium (Y), aluminum (A), and garnet (G) with a few percent of neodymium (Nd) uniformly dispersed within. Although the Nd:YAG (synthetic) laser is one of the most popular types on the market, the risk of potential eye damage is greatest because of this wavelength. The 1.06 μm wavelength is in the ocular focus region (0.35-1.40 μm) so that the radiation intensity on the cornea is amplified 10^5 times by the ocular system and focused onto the retina. Also, this wavelength is *not visible* by the eye—it is *invisible*. In addition, retinal damage threshold values are quite low, especially for pulsed YAG lasers, on the order of 10 microjoules per square centimeter (μJ/cm^2) incident on the cornea for 30 picosecond (30 ps) pulses (2). Control of laser hazards is discussed in a later chapter.

The Lawrence Livermore (CA) National Laboratory has developed the world's largest Nd:glass laser system, capable of producing over 100 trillion watts of optical power in a burst of laser light (1.06 μm wavelength) lasting one-billionth of a second (see Fig. 2.2).

In the past few years, research with a variety of new solid-state materials has produced not only new lasers, some of which are "tunable" (capable of emitting several specific individual wavelengths), but has also improved the efficiency of the previous solid-state lasers, even producing new wavelengths in the visible and infrared regions. Chromium and titanium lasers are two examples of new tunable lasers. The field of tunable solid-state lasers has developed so rapidly that the Optical Society of America and the IEEE (Lasers and Electro-Optics Society) cosponsored a second Topical Meeting on Tunable Solid-State Lasers in June 1986 that featured papers on (a) chromium and titanium lasers, (b) new transition metal lasers, (c) color center lasers, (d) new rare earth lasers, and (e) tunable laser applications.

SEMICONDUCTOR LASERS

A semiconductor laser consists of a flat junction of two pieces of semiconductor materials, each of which has been treated with a different type of impurity. When an electrical current is passed through such a device, laser light emerges from the

Figure 2.2 The world's largest Nd:glass laser system, "Nova." (Courtesy of Lawrence Livermore National Laboratory, Lawrence, California.)

junction region. Power output is limited, but the low cost, small size, and relatively high efficiency make these lasers well suited for microelectronic adaptation, and applications are almost unlimited in high-technological uses, for example, fiber optics communications. These lasers are similar in construction to a transistor or a semiconductor diode. They are usually infrared pulses with power on the order of watts and can be produced with good efficiency, especially at low temperatures, liquid nitrogen or lower. The activation mechanism can be a voice or television current signal, thereby producing laser beams modulated with these signals.

GAS LASERS

Semiconductor materials such as silicon, conduct electricity better than insulators but not as well as true conductors. These semiconductor materials allow construction of complex microelectronic circuits by carefully controlling the composition making it possible to build many useful structures to accommodate the desired use.

Over the past few years, the semiconductor laser technology has been one of the fastest growing of all the laser types because they are essential in the fiber optics communications field wherein the need exists to "relay" weakened signals for long distance transmission.

The first practical commercial semiconductor laser was available in 1975, and advances made in this field since then have resulted in room-temperature lifetimes of about 100 years for continuously emitting semiconductor lasers. Some projections of future lifetimes include millions of years of useful lifetimes. Advances are being made toward producing beams of high quality and low diversion angles. Research with new materials, such as indium and phosphorus, in addition to the standard gallium, arsenic, and aluminum (which emit wavelengths of 0.8-0.9 μm in the infrared), has resulted in longer wavelengths of 1.1-1.6 μm, which increases efficiency of light transmission in optical fibers. Also being developed are shorter wavelengths, such as in the visible (red, particularly) and, even in the ultraviolet (<0.4 μm), but these are research oriented at this writing.

GAS LASERS

Gas lasers produce conherent light beams by an electrical discharge in the laser cavity or gas chamber. Several types of gas lasers have gained popularity because of (a) the simplicity of the lasing function, (b) the wide variety of wavelengths available, (c) they are less expensive, (d) some, for example the CO_2, are very efficient (~30%), and (e) very high power outputs are available from relatively small units.

A simplified diagrammatic sketch is shown in Figure 2.3 that explains (a) the gas cavity, (b) excitation of the gas by electrical charge, and (c) stimulated emission of the laser light on discharge of the electrical pulse. A variety of "pumping" methods are used in gas laser technology whereby the atoms or molecules of the lasing gas are excited. In some gas mixtures, such as the most popular of all, helium-neon, the electrical discharge excites the helium atoms which transfer their energy to the neon atoms which emit red light. Other gas mixtures, such as carbon dioxide (CO_2), nitrogen, and argon, are energized by the CO_2 molecules absorbing the electrical energy and vibrating at a high energy level.

The helium-neon laser cannot produce much power, typically only a few milliwatts, but this versatile laser can emit laser light continuously for many thousands of hours and has a plethora of uses (see Fig. 2.4).

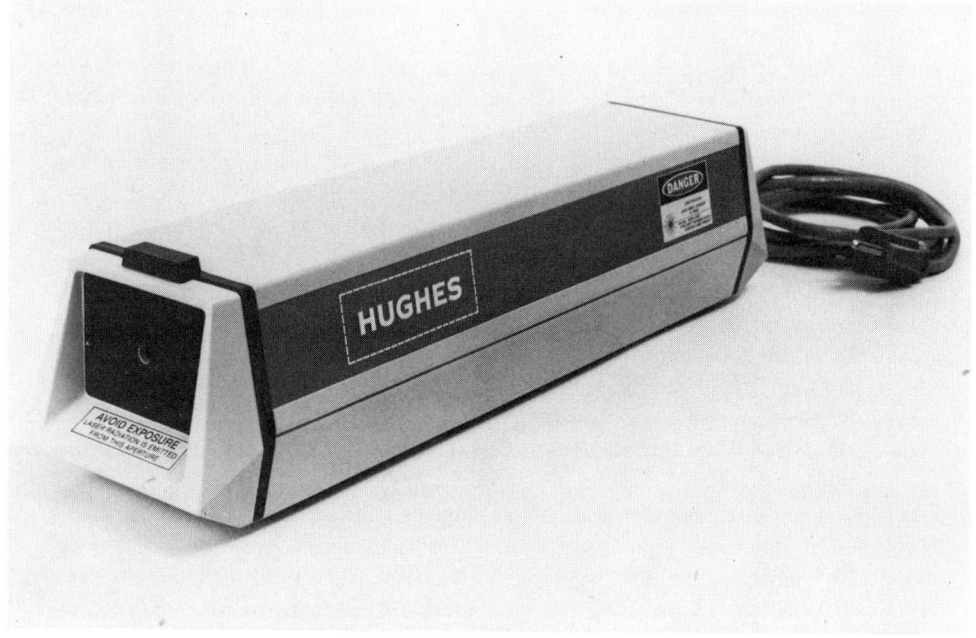

Figure 2.3 Current design of nominal 5 mW output of helium-neon gas laser featuring the laser head and power supply contained in a single housing. (Courtesy of Hughes Aircraft Company.)

In laser fusion research at the Los Alamos National Laboratory, the CO_2 laser has been developed to produce terawatts (10^{12} W) in pulses of ultrashort pulsewidths (10^{-12} sec). Amplification is possible by making multiple passes of the beam through the huge gas cavities developed to produce clusters of beams of several feet in diameter. Figure 2.5 shows a huge amplifier gas chamber of Antares, the world's largest CO_2 laser system. Figure 2.6 shows the power supply board from a 3J CO_2 laser developed for the Antares project.

Continuous beams of CO_2 lasers of high intensity are also available for such applications as welding or cutting metals.

Other popular gas lasers are the argon and crypton lasers which emit in the green and red portions of the spectrum, respectively. Mixed together, the resulting emissions can be controlled to provide four individual and separate wavelengths in the visible range. One application popular with this combination is in the entertainment field—light shows, movies, etc.

CHEMICAL LASERS

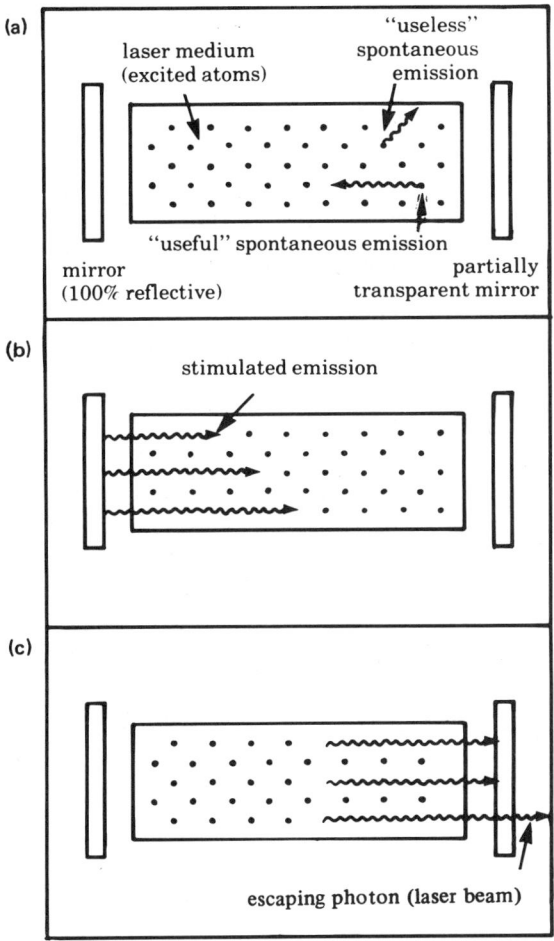

Figure 2.4 Diagrammatic sketch showing (a) excitation of gaseous laser medium (i.e., electrical discharge), (b) reflection of photons during stimulated emission, and (c) exiting photons in a laser beam.

CHEMICAL LASERS

Certain chemical reactions produce enough energy to excite the materials sufficiently to produce laser activity on formation of stable products. The most important of these reactions occurs when the elements hydrogen and fluorine are ignited to form hydrogen fluoride (HF). The initial product of the reaction is a

Figure 2.5 A view of a technician on the space inside a dual-pass amplifier of Antares, the world's largest carbon dioxide laser at the Los Alamos National Laboratory. The facility is used in studying basic physics of fusion reactions on a laboratory scale. Notice large diameter electrical leads that supply the electrons needed for activation of the huge volume of gas (including CO_2) required to amplify the original beam. (Courtesy of Los Alamos National Laboratory, Los Alamos, New Mexico.)

molecule of HF in an excited state of vibration. If an appropriate laser cavity is provided for this reaction, the energy from the excited molecules can be extracted as laser light. In practice, this method of providing an infrared laser beam of 3 μm wavelength includes flowing the gases into the cavity where a continuous flow produces a continuous beam. Pulsed HF lasers are made possible by igniting controlled quantities of the gas mixture. Pulsed HF lasers are attractive in some applications such as inertial confinement fusion research and in military applications, because extremely high power is possible without requiring cumbersome electrical energy power supplies for excitation. However, both hydrogen and

LIQUID LASERS

Figure 2.6 Photograph of a technician with a power supply board from a 3J CO_2 laser originally developed for the Antares system. (Courtesy of Pulsed Systems, Los Alamos, New Mexico.)

fluorine are potentially hazardous materials and extreme care must be exercised in their use (2). Commercial HF lasers are not very popular for these reasons, but research is continuing in this field, and perhaps other candidates, albeit less potentially hazardous, may be developed to produce laser beams by this simple process of producing energy by chemical reaction.

LIQUID LASERS

Several types of liquids have been used to create laser beams, but only one kind has been developed into a device of general usefulness. While a small number of inorganic liquids can function as lasers, the organic dye laser is the most popular. It consists of solid dye materials dissolved in an organic solvent such as alcohol to form a solution. Organic dyes have energy levels so closely spaced that a number of these levels can produce a wide range of wavelengths in the visible portion

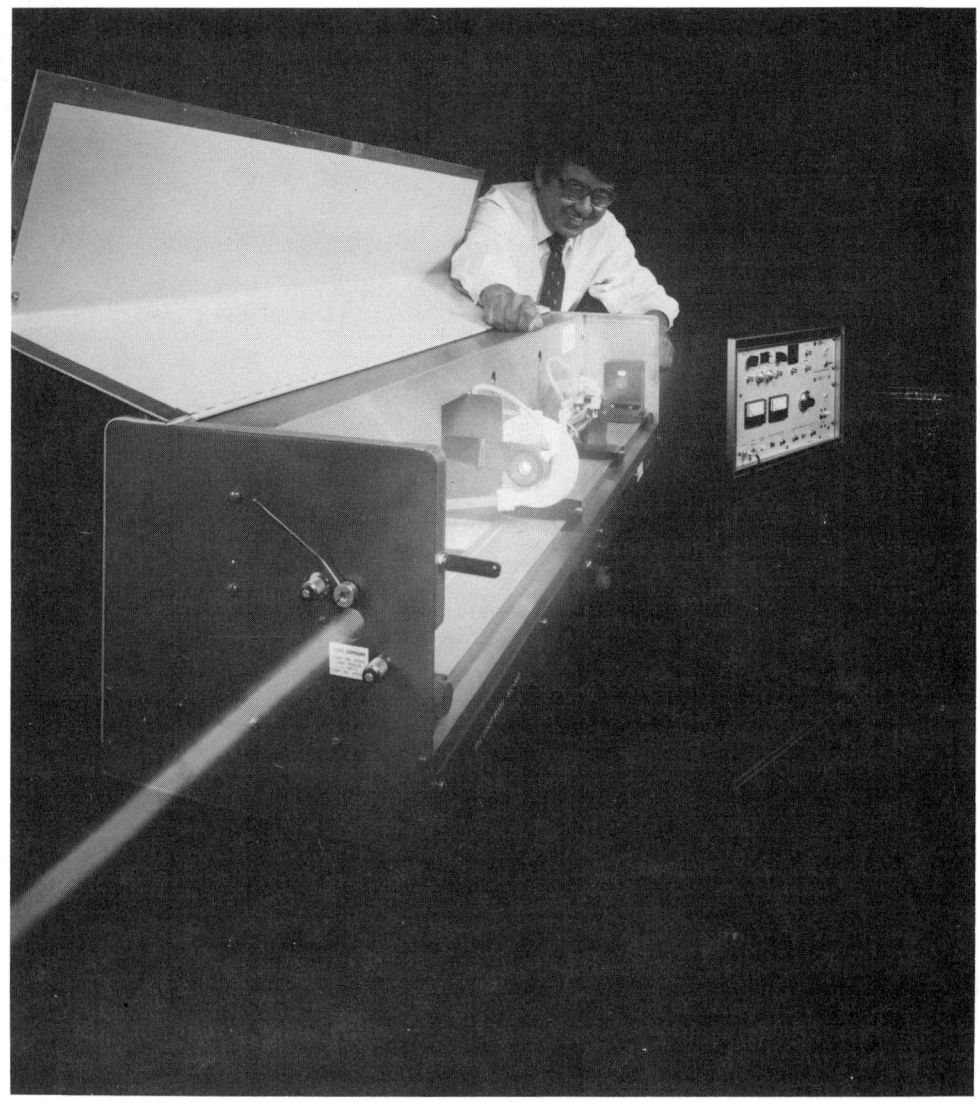

Figure 2.7 Photograph of a tunable laser used mainly in research or applications requiring the utilization of a variety of wavelengths within specific ranges. (Courtesy of Candela Corporation, Natick, MA.)

of the spectrum, 0.4-0.7 μm. Using an appropriate optical arrangement, a single wavelength emitting from the dye solution can be selected for use in the process at hand. For example, these lasers are used in applications involving chemical analysis and identification of atomic species. Because of the wide range of wavelengths available in a variety of dye solutions, dye lasers are described as "tunable" (see Fig. 2.7).

Solid lasers may suffer damage at high power levels due to the intense heat generated within the solid material and by the pumping lamp. The liquid laser is not susceptible to such damage. The cavity contains the solution in a transparent cell and excitation is performed by another laser. A dye identified as Rhodanine 6G was the first dye for which continuous, rather than pulsed, operation was achieved, making possible the production of a continuous beam of tunable laser light. Another dye, methylumbeliferone, with the addition of HCl, can be made to lase at wavelengths varying across the light spectrum from ultraviolet to yellow, producing laser light of almost any desired frequency within this range.

According to Hecht and Teresi (1) one of the strangest lasers ever produced was a dye laser. It was a disk-shaped laser that emitted light along all 360 degrees of its circumference. In this device, the circumference was coated with partially reflecting film, producing a disk-shaped laser cavity. At the center of the laser was a dye that was pumped by another laser. One of the inventors believes it should be possible to extend his technique to make a spherical laser, one that would emit laser light from the entire surface of a ball-shaped device.

EXCIMER LASERS

Although excimer lasers are of the gas type, special attention is given to them because they also could be considered a chemical laser. Essentially consisting of a mixture of a rare gas (helium, argon, krypton, neon) and a halogen (chlorine, fluorine, bromine, iodine) in a cavity into which energy is deposited by an electron beam or an electrical discharge to cause an electrically excited molecule which can only exist in this excited state. The excimer, for example the currently most popular, krypton fluoride (KrFl), immediately returns to its constituent atoms emitting photons, either spontaneously or by stimulated emission to produce a laser beam at 0.250 μm, in the ultraviolet region (.1-.4 μm) of the spectrum.

Because excimer lasers are gas and chemical types and because the wavelength is shorter than the other popular types, it is of interest for many applications. Since amplification is rather straightforward—build bigger cavities, and because shorter wavelengths of high intensity might "couple" better with the inertial confinement fusion targets, the excimer-type laser is useful in fusion experiments. Also, since the short wavelength reacts on the outer surfaces of target materials, the medical field has found applications for its use in surgical procedures.

FREE ELECTRON LASERS

According to an article in the September 13, 1985, edition of the *Los Alamos Newsbulletin*, the official organ of the Los Alamos National Laboratory, "The Laboratory's Free Electron Laser has emerged as the most tunable in the world— able to successfully test-fire 6,000 watt beams of light at a variety of wavelengths spanning the infrared portion of the spectrum." Jeff Schwartz, author of the article, adds the following description of how the laser works.

"Free electron lasers (FELs) represent a totally new concept in the quest to develop powerful light sources for strategic defense, industry and basic research applications.

They produce light beams with the aid of electrons, the same subatomic particles that travel through household wiring to meet daily electricity needs.

In the case of the FEL, the electrons are generated by accelerators and then directed through an array of 340 magnets, 3 feet in length, arranged with alternating north-south poles.

The negatively charged electrons passing through the array are then forced to 'wiggle' as the alternating magnetic poles deflect these tiny particles left and right. This wiggling causes the electrons to emit light, a phenomenon well-known to physicists.

Light is reflected back and forth between two small mirrors spaced 23 feet apart, gaining strength on each pass through the wiggler magnets. Each round trip between mirrors takes about 46 billionths of a second During a test shot, 2,000 trips are made.

FEL scientists can literally "dial" a wavelength." The wavelength produced by the free electrons is related to the particles' energy. As a result, scientists are able to dial up a specific wavelength by choosing an accelerator known to supply electrons at energy levels that radiate certain laser wavelengths when wiggled. This is a key advantage that FELs have over conventional lasers that emit only a single wavelength beam when their medium is excited to produce light.

Another attractive characteristic of such lasers is the high-quality beam emitted by free electrons when they move through a magnetic field. For example, the beam from a spotlight, consisting of light from a variety of wavelengths, would be spread about 2,000 miles across by the time it reached the moon.

By comparison, a laser beam of perfect optical quality, such as an FEL, would be only 200 yards in diameter at the same spot because it is made of only one specific wavelength, and the beam divergence is low.

The first free electron laser was tested at Stanford University in 1976. Since then, Los Alamos researchers have been world leaders in this field.

"Los Alamos was the first to demonstrate high power, the highest efficiency and now the broadest tunability," said FEL Program Manager Charles Brau.

In the future, attempts will be made to tune the FEL down to very short wavelengths, such as reaching light in the ultraviolet and x-ray spectrum.

Damage to mirror coatings will be a problem with such beams, but the difficulty of producing short wavelengths using other techniques makes the FEL an approach to be explored.

In the meantime, an energy recovery system is being added to improve the FEL's efficiency.

The system will recycle the electron beam after it leaves the magnetic wiggler to get more laser energy for the same input power, thus lowering both capital and operating costs.

Scientists have now begun to consider potential applications as free electron lasers demonstrate success in research. They could prove useful to the semiconductor and medical industries and to chemistry research.

In addition, high-power beams with good optical quality might make the FEL a possible "star wars" candidate for the future, since these lasers could be either "ground- or space-based."

For additional information on the free-electron laser see Chap. 7, Other Medical Applications.

OTHER LASER TYPES

Metal Vapor Lasers

Metal vapor laser technology is in its infancy, but these lasers are mentioned here because of their potential beneficial effects in cancer treatment. Photodynamic therapy (PDT) is a technique of killing cancer cells by allowing a "hematoporphyrin derivative" (HpD) to accumulate in the tumor bed and then irradiating it with a laser beam of appropriate wavelength, releasing free radicals of oxygen from the HpD to attack cancer cells. Gold and copper lasers or even a mixture of these metals may provide the appropriate wavelength in the PDT applications. Tunability is an important quality in this type of research, therefore dye lasers are of interest here also.

Design of metal vapor lasers is similar to other gaseous laser systems, having an electrical discharge facility and mirrors at the ends of the cavity. This arrangement makes the metal vapor laser economically competitive. Several manufacturers offer these lasers for research and development applications.

X-Ray Laser

Physicists are considering the taming of x-rays to perform as a coherent beam source. Early laboratory experiments have been in the wavelength range of 0.01-0.02 μm, or the "soft" x-ray region of the spectrum.

The effort is described by Rosen et al. (3) as follows: "... A 0.53-μm laser, focused on a 1.2 × 0.02-cm spot to 5 × 10^{13} W/cm^2, heats and burns through a thin foil of selenium. Besides ionizing the selenium to a neon-like state, the laser explodes the foil, creating a region of uniform electron density. This allows propogation of the x-rays down a 1-cm-long gain direction without debilitating refraction. Gains of 4 to 10 per cm are predicted for various transitions."

As in the early stages of other laser development, applications for the x-ray laser are unknown, except that availability of this wavelength opens up new research vistas. Also, on a grand scale, applications are anticipated in the "star wars" program as a possible defensive deterrent or shield for enemy nuclear weapon intrusion.

Miscellaneous Lasers

Many new materials are being considered for laser cavities. These include a chemically pumped oxygen laser and, if successful, would be emitting in the visible (red) portion of the spectrum. It is being touted as having possible applications in defense and power transmission in space as well as having possible material processing functions. A report on work in the Soviet Union concerns an optically pumped mercuric halide laser emitting in the blue-green wavelengths where radiation penetrates sea water. Exotic approaches to researching lasing techniques include investigating color-center laser potential. This is truly the age of the laser, and every field of endeavor must determine its usefulness in making that technology as useful and efficient as possible in modern society.

REFERENCES

1. Hecht, J. and D. Teresi (1982). *Laser, Supertool of the 1980's*. Ticknor and Fields, New York.
2. Winburn, D.C. (1985). *Practical Laser Safety*. Marcel Dekker, New York.
3. Rosen, M.D. et al. (1985). Exploding-foil technique for achieving a soft x-ray laser. *Physical Review Letters*, January 14, 1985.

3
Characteristics of Laser Beams

Light has a dual nature. In travelling through space, or in being reflected, it behaves like a wave disturbance. In the process of being emitted from, or absorbed by, material substances, light behaves like a particle. In this chapter, extracted from DeMommio et al. (Ref. 1), the wave properties of light will be examined. For some engineers with a working knowledge of optics this chapter will be a review of fundamental physics, but it is intended to serve as a precursor to the more technical data presented in Chapter 4. Another reference for fundamentals of optics is Jenkens and White (2).

ELECTROMAGNETIC SPECTRUM

Many natural phenomena cause periodic disturbances that are represented by waves that are repetitious, or periodic, such as shock waves from an earthquake, ripples on still water when a pebble is dropped in a pool, or sound waves from a musical instrument. Light propagation can be represented by waves also. All wave phenomena exhibit one common repetitive property—the wave is cyclic, that is, the wave repeats itself a certain number of times per unit length of time. The number of times the wave repeats itself is called the "frequency" of the wave. This characteristic of a wave can be represented graphically, together with other wave functions, such as "amplitude," which indicates the distance travelled from

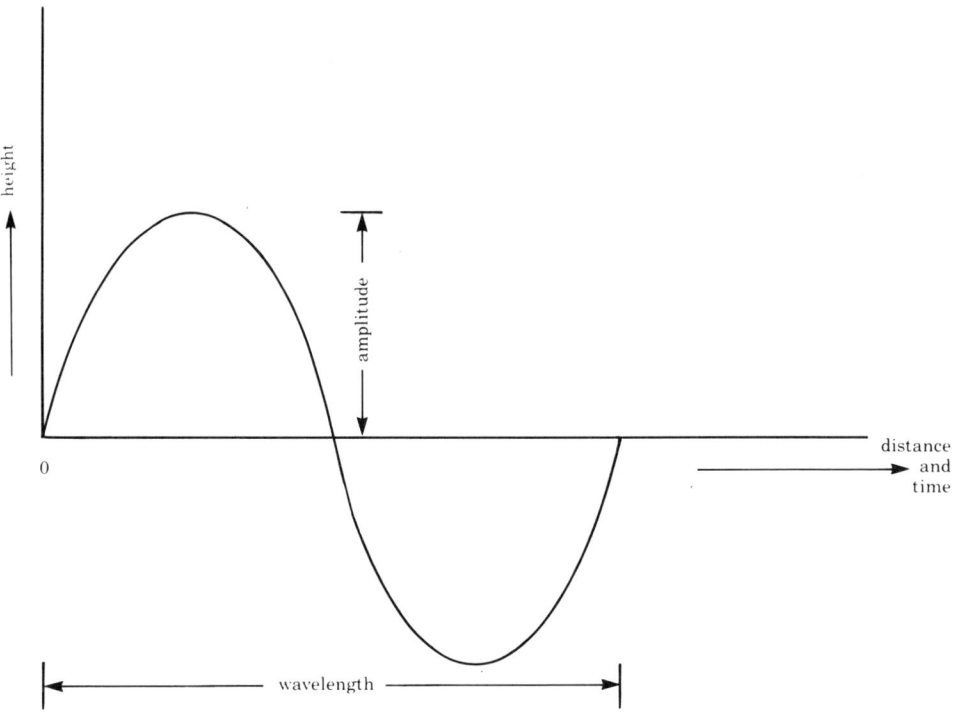

Figure 3.1 Wave properties. The wave, starting at zero, reaches a maximum amplitude, returns to zero, repeats its amplitude in the opposite direction, and returns to zero for a complete cycle, or period of time, or one wavelength in distance.

zero to a maximum value and the "wavelength," as shown in Figure 3.1. The distance travelled in one cycle of time or period is called the wavelength and is represented in all light wave calculations as lambda, λ. Each light emitter, such as a laser, has a specific wavelength. Because the speed of light, c, is known to be 3×10^8 meters per second, the frequency, v, or wave repetition rate, or wavelength, can be calculated from the formula

$$c = v\lambda$$

Consequently, if either the frequency or wavelength of a laser beam is known, the unknown characteristic can be calculated. For example, if the common Nd:YAG laser emits a beam of 1.06×10^{-6} meters (1 μm, or micron) in wavelength, the frequency could be calculated:

$$c = v\lambda \quad \text{or} \quad v = \frac{c}{\lambda} = \frac{3 \times 10^8 \text{ m/s}}{1.06 \times 10^{-6} \text{ m}} \quad \text{or}$$

ELECTROMAGNETIC SPECTRUM

the frequency, $v = 2.83 \times 10^{14}$ Hertz. Now, if the frequency of this laser is doubled,* as is a popular device for obtaining shorter wavelengths, the wavelength is thereby halved to 0.53 μm.

At this point, representations of the electromagnetic spectrum are inserted as Figures 3.2 and 3.3 to use in following a discussion of laser wavelengths, colors, and frequencies. Also of interest is the table in Figure 3.4 showing abbreviations or symbols of units and metric prefixes with exponential factors.

A model of an electromagnetic light wave, representing a laser wavelength, moving through space is shown in Figure 3.5. The discussion of this model was

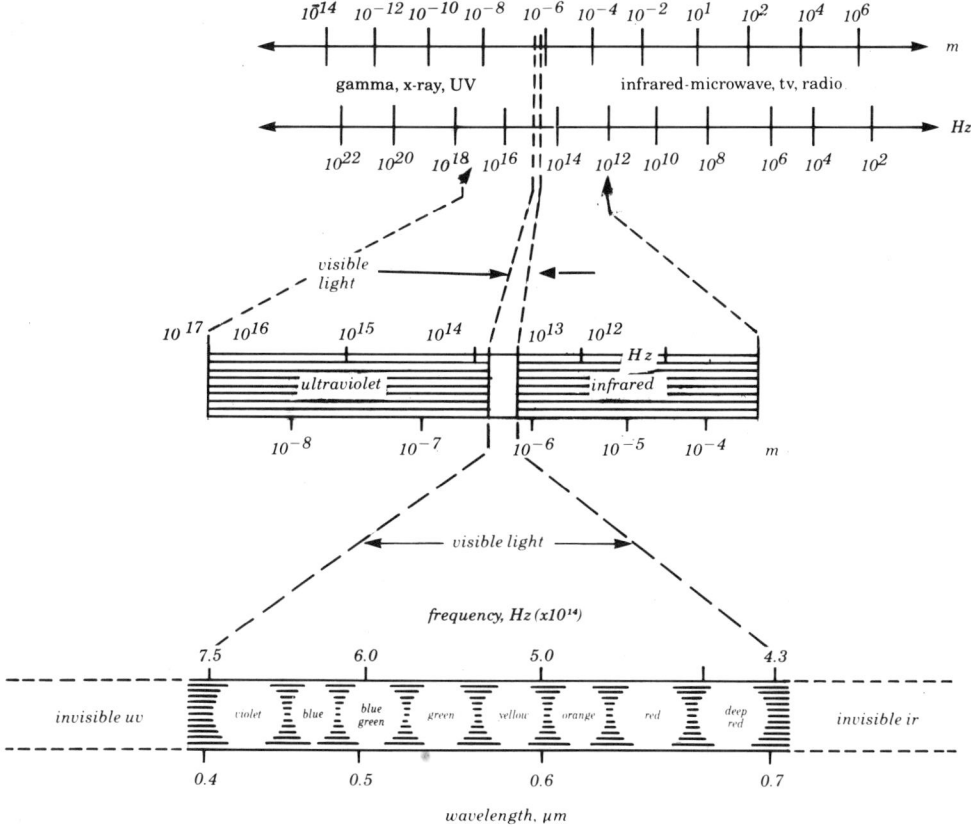

Figure 3.2 The electromagnetic spectrum (E-M).

---

*A process easily obtained by passing the 1.06 μm beam through a special crystal.

CHARACTERISTICS OF LASER BEAMS

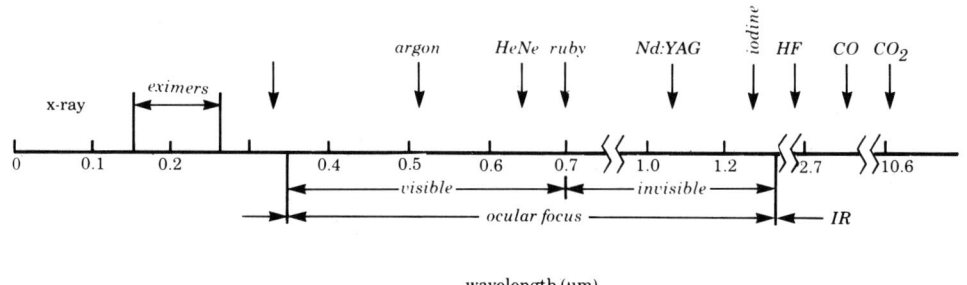

Figure 3.3 Location of common lasers in electromagnetic spectrum.

Prefix	Symbol	Factor
ultra*	U	10^{15}
tera	T	10^{12}
giga	G	10^{9}
mega	M	10^{6}
kilo	k	10^{3}
hecto	h	10^{2}
deka	da	10^{1}
--	--	10^{0} (or unity, 1)
deci	d	10^{-1}
centi	c	10^{-2}
milli	m	10^{-3}
micro	μ	10^{-6}
nano	n	10^{-9}
pico	p	10^{-12}
femto	f	10^{-15}
atto	a	10^{-18}

*Suggested by author.

Examples: 1 km = 10^{3} m = 1000 m
1 mm = 10^{-3} m = 0.001 m

Figure 3.4 Metric prefixes.

ELECTROMAGNETIC SPECTRUM

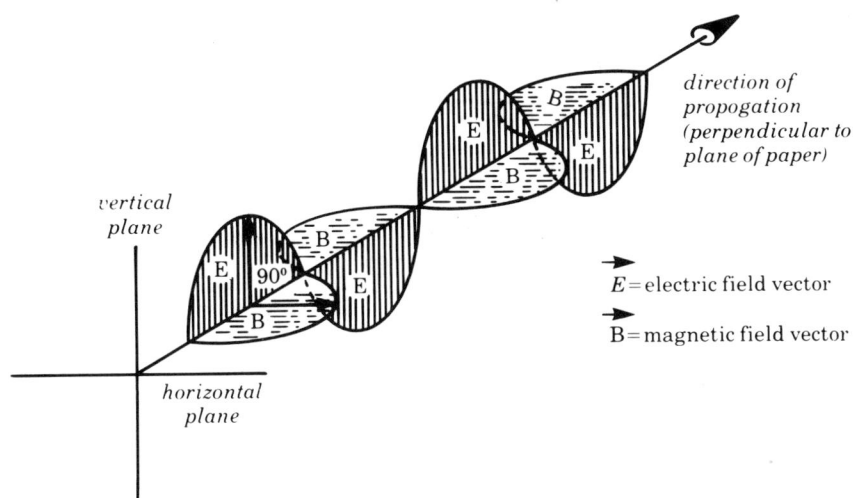

Figure 3.5 Spherical and plane wavefronts.

developed by the Technical Education Research Centers, Inc., in a 1972 publicationdescribed as "Laser Electro Optics, Course 1, Introduction to Lasers" (1).

The E-M wave shown in Figure 3.5 is an example of a transverse wave. All E-M waves consist of two waves, one electric and the other magnetic in character, perpendicular to one another, moving through space at a certain speed. The amplitude of a transverse wave is perpendicular to its direction of propagation. The E→ and B→ vectors in Figure 3.5 are both perpendicular to the direction of travel of the wave as well as perpendicular to each other. Electromagnetic waves include radio waves, television, microwaves, x-rays, gamma rays, as well as the narrow band of frequencies we call "light." Laser beam frequencies exceed the extremities of both the ultraviolet (UV) and infrared (IR) regions. *All E-M waves have the same basic nature and properties—they differ only in frequency.* The truly enormous span of frequencies covered by electromagnetic phenomena is shown in Figure 3.2.

This collection of frequencies and E-M wave phenomena is known as the *electromagnetic spectrum.* It must be emphasized that it is the frequency of a particular E-M wave that gives rise to a given type of generating or transmitting device and corresponding receiver or detector. Hence radio frequency transmitters and receivers, having frequencies of 10^3-10^6 Hz, use quite different components than generators and detectors of light waves, having frequencies of the order of 10^{14} Hz.

Electromagnetic waves of all frequencies in the spectrum are found experimentally to travel at the same speed, c, in a volume, where $c \cong 3 \times 10^8$ meters/sec.

Velocity of a wave, c, has the following relationship to frequency and wavelength:

velocity = (frequency) (wavelength)   or $c = v\lambda$

where v = frequency (Hz) and, λ = wavelength (meters). For example, green light from an argon gas laser has a wavelength λ = 514.5 nm, where 1 nanometer (nm) = 10^{-9} meter. What is the frequency of the waves emitted by such a laser? A simple calculation shows:

$$v = \frac{c}{\lambda} = \frac{3 \times 10^8 \text{ m/sec}}{5.145 \times 10^{-7} \text{ m}}$$

$$= 5.8 \times 10^{14} \text{ Hz}$$

This means that the electric (or magnetic) field vector in the light beam is oscillating at the rate of some 100 trillion times each second.

What makes E-M waves so useful? All types of waves, whether they are of mechanical or electromagentic origin, *carry energy*. The transport and fine degree of control of energy over some distance is of great importance for many applications. Electromagnetic waves transport energy through space from the point of origin (the source of the "disturbance") to some receiving point where the energy may be collected and used. Figure 3.5 indicates how the energy is radiated in an E-M wave.

The spatial nature of a beam radiated by a source of E-M waves can be represented by a series of surfaces joining points of maximum vibration (wave crests) and minimum vibration (wave troughs). These surfaces are special examples of surfaces called *wavefronts*. If the source is effectively an infinitesimal point from which waves are emitted, then *spherical* wavefronts are produced, as shown in Figure 3.6. Spherical wavefronts may be approximated by light radiated by a tiny light bulb. *Plane* wavefronts are also shown. The beam emitted by a CW (continuous wave) gas laser forms wavefronts that are nearly planar.

To obtain further insight into wavefronts, we need to introduce the concept of *phase*—another property of wave motion. In general, a *wavefront* is defined as a *surface* of *constant phase*. To avoid complex mathematics and without sacrificing any of the fundamental principles involved, we can restrict the discussion to the case of a plane electromagentic wave characterized by a single frequency and polarization. When this is done, it can be shown that the amplitude of the wave in either time or space is given by the sinusoidal form shown in Figure 3.7. The sinusoidal form can be related to a circle as shown and for this reason it is convenient when discussing sinusoidal waves to measure the horizontal position (horizontal in Fig. 3.6) along the wave in terms of degrees or radians with 360 degrees or 2π radians representing the horizontal displacement for a complete cycle. In our discussion so far we have treated electromagnetic waves

ELECTROMAGNETIC SPECTRUM

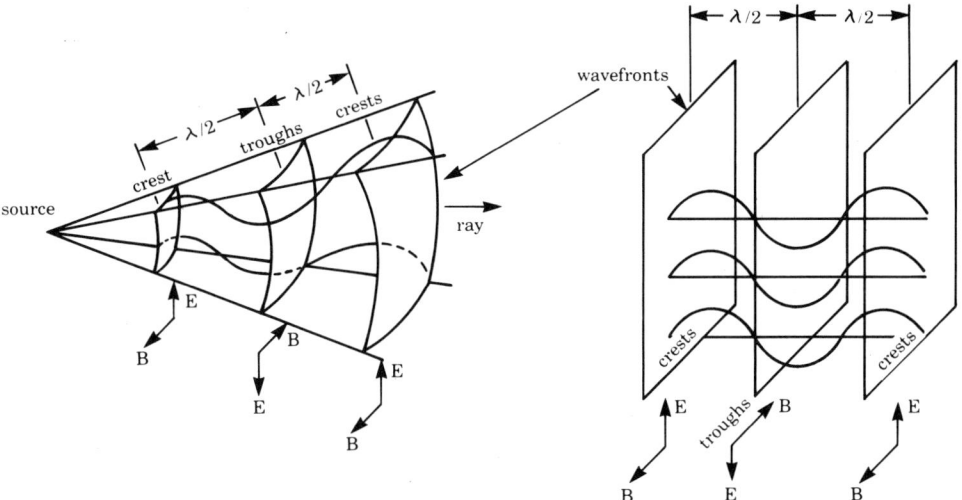

Figure 3.6 Example of a phase angle.

as though they have zero amplitude at zero time. There is, of course, no reason why this has to be so. We can still completely define the wave if at time t = 0, we give its amplitude relative to its peak amplitude and also state whether it is increasing or decreasing. We can accomplish the same thing more simply by stating how far in time or position the point on the wave in question is horizontally from the last positive-going zero crossing. The value of this distance or time is called the *phase* of the wave.

For example, in Figure 3.7, the double circle in the drawing indicates the position from which oscillations began at time t = 0. In this example, the phase angle $a = 5\pi/6$ radians or 150°. Notice that the displacement of this particular wave at t=0 has a non-zero value. We can see now that a wavefront, when defined as a surface of constant phase, means that surface obtained when all of the points on an electromagnetic wave having the same phase with respect to a particular zero crossing are connected. The reference zero crossing points when so connected are, of course, a special case of such a wave front.

Two waves are said to be *in phase* if their respective phase and angles are equal, and *out* of *phase* if their phase angles are not equal. For example, if wave 1 has a phase angle $a_1 = \pi$ radians (180°), and if wave 2 has a phase angle $a_2 = 3\pi/2$ radians (270°), then these two waves are out of phase with one another by an amount $\Delta\phi$, where $\Delta\phi = (a_2 - a_1) = 90°$ or $\pi/2$ radians.

Polarization refers to the orientation of **E** and **B** field vectors in space. If the tip of the electric field vector **E** oscillates along a line in a plane perpendicular

28 CHARACTERISTICS OF LASER BEAMS

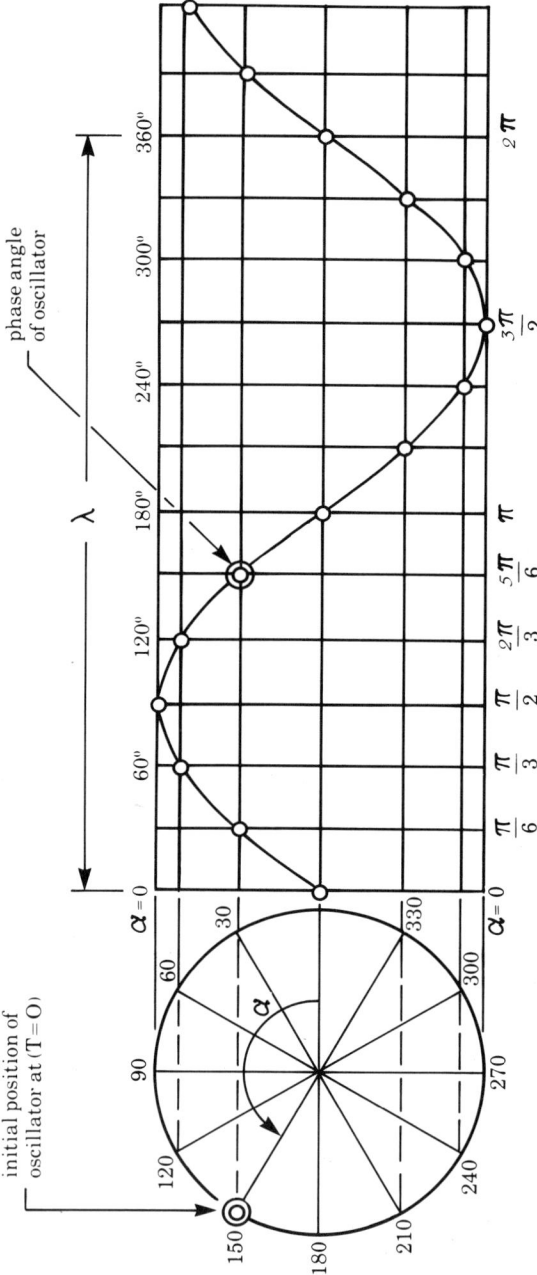

Figure 3.7 Three-dimensional model of a plane-polarized E-M wave.

to the direction of wave propagation, the wave is said to be *linearly polarized*. The model of an E-M wave shown in Figure 3.7 is a specific example. If the tip of the electric field vector **E** traces out a circle in a place perpendicular to the direction of wave propagation, it is said to be *circularly polarized*. If the tip of the electric field vector moves randomly in the plane perpendicular to the direction of wave propagation, the wave is said to be *unpolarized*.

Laser beams are often linearly polarized. This can be tested by inserting a device called a linear polarizer in the beam path. A linear polarizer has the effect of passing light whose **E** vector is parallel to a certain direction called the *transmission axis*. The effect of a linear polarizer on an unpolarized beam is to separate by transmitting only those electric field components parallel to the transmission axis.

A traveling E-M wave contains energy. The power density, E, of an electromagnetic wave is defined as the average power, P_{av}, in the beam divided by the area, A_b, of the beam:

$$E = \frac{P_{av}}{A_b}$$

It has units of watts per square meter (W/m^2).

The power density is proportional to the square of the amplitude of the electric field in the E-M wave:

$$E = (1.33 \times 10^{-3} \text{ W/V}^2)\Sigma^2,$$

thus we can use a linear polarizer to produce linearly polarized light from an unpolarized beam, as shown in Figure 3.8. In addition, we can test a beam to see if it is already linearly polarized by inserting a linear polarizer in the beam path and rotating it through an angle of 90°. Refer to Figures 3.8 and 3.9. If the original beam is linearly polarized, we should be able to cancel or nearly cancel it by orienting the linear polarizer so that its transmission axis is perpendicular to the polarization plane of the incident light beam. A measure of how well a beam is linearly polarized is known as the *extinction ratio* (ER). This is the ratio of the power transmitted through the polarizer when its transmission axis is aligned with the polarization plane (maximum power) to the power transmitted when the polarizer is rotated 90° (minimum power):

$$ER = \frac{P_{max}}{P_{min}}$$

where Σ is the peak amplitude of the electric field in volts/meter. The numerical constant simply insures that E will be in W/m^2 if Σ is in volts/meter. Thus, if we measure the power density, E, of a laser beam, we can determine the amplitude of the electric field. For example, consider a repetitively pulsed neodymium:YAG laser of average output power P_{av}=5 kW and beam diameter d=7 mm. The beam power density, E, or intensity, is calculated

30   CHARACTERISTICS OF LASER BEAMS

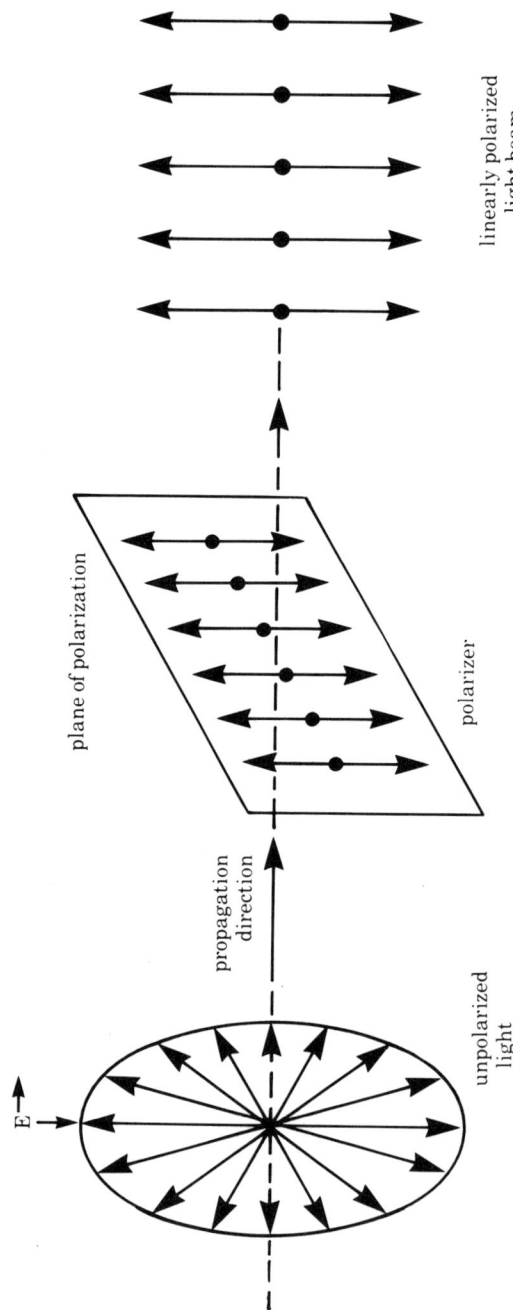

Figure 3.8 Producing linearly polarized light.

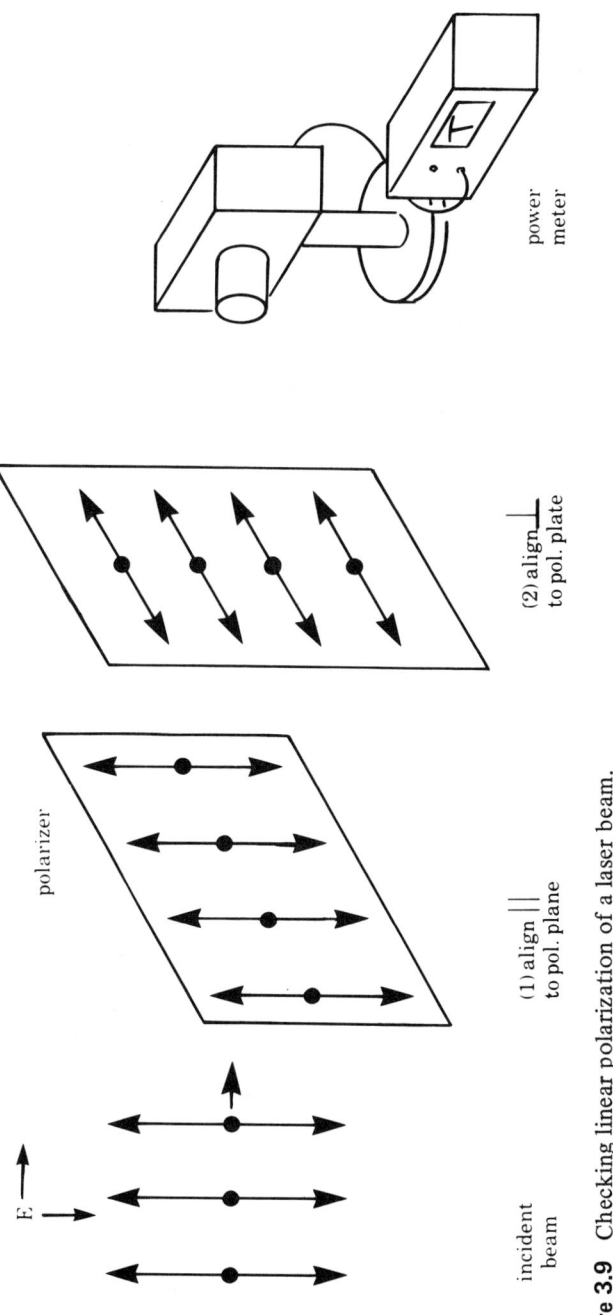

Figure 3.9 Checking linear polarization of a laser beam.

$$E = \frac{P_{av}}{\pi(d/2)^2}$$

$$E = \frac{5 \times 10^3 \text{ W}}{3.14 \, (3.5 \times 10^{-3} \text{ m})^2}$$

$$E = 1.30 \times 10^8 \text{ W/m}^2$$

The amplitude of the electric field, Σ, may be obtained

$$\Sigma^2 = \frac{1}{1.33 \times 10^{-3} \text{ W/V}^2}$$

$$\Sigma^2 = \frac{1.30 \times 10^8 \text{ W/m}^2}{1.33 \times 10^{-3} \text{ W/V}^2}$$

$$\Sigma = 3.13 \times 10^5 \text{ V/m}$$

A rather convincing experiment on the wave nature of light can be performed with two pieces of flat glass forming a wedge. Examine Figure 3.10. If a monochromatic light source, such as a laser beam, illuminates the wedge, a pattern of bright and dark bands can be seen. These are called *fringes*. This effect

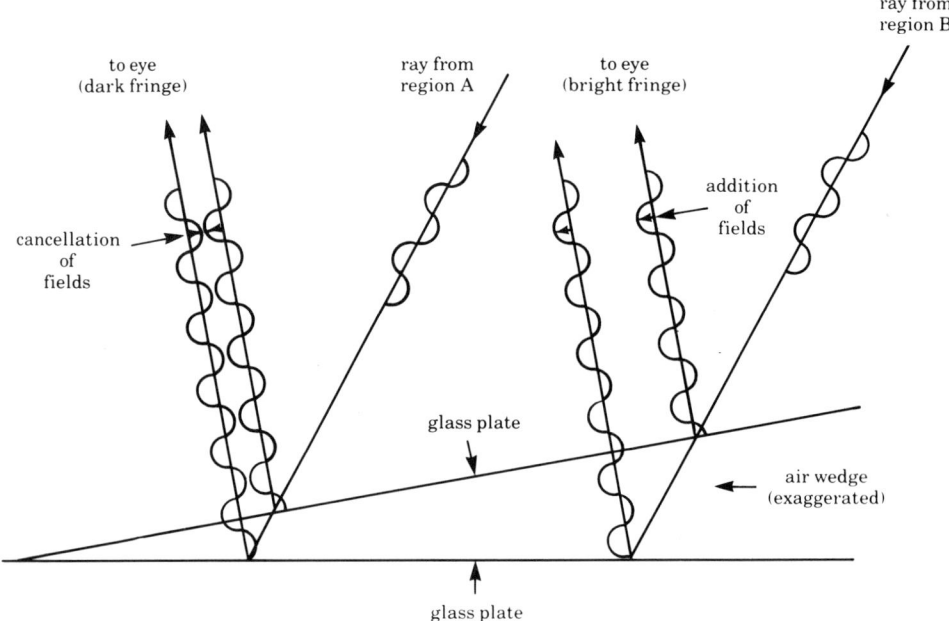

Figure 3.10 Interference of light waves.

WAVELENGTH EFFECTS

can be explained if we assume that light behaves, in part, like a wave. The bright fringes correspond to *constructive* interference of light waves, the dark fringes to *destructive interference*. The first case corresponds to an interaction of two waves that are *in phase* with one another, the latter to two waves that are *out of phase* with one another by 180°. The equations which follow give the conditions for interference when viewed nearly perpendicular to the surface:

<u>Destructive interference</u>: $2t = m\lambda$, $(m = 0, 1, 2, \ldots)$
(dark fringes)

<u>Constructive interference</u>: $2t = (m + \frac{1}{2})\lambda$, $(m = 0, 1, 2, \ldots)$

In these equations, t is the air wedge thickness. For example, suppose a pair of flat glass plates, separated by a small air wedge, is illuminated with monochromatic light from a sodium arc lamp of wavelength $\lambda = 589$ nm. The minimum thickness of the thin film of air between the plates for which constructive interference of light waves will occur, according to the constructive interference, equation with m=0, is

$$2t = \frac{\lambda}{2}$$

$$t = \frac{\lambda}{4}$$

$$t = \frac{589 \text{ nm}}{4}$$

$$t \cong 147 \text{ nm}$$

This thickness is about 58 ten-thousandths of an inch, 0.0058".

WAVELENGTH EFFECTS

The first consideration in selecting a laser to perform a specific function is to determine the wavelength desired. Evaluation of this prime characteristic must precede procuring any device, because each wavelength has its own unique properties and a match of the need or use must be established initially. A cursory knowledge of such characteristics as transmission of certain wavelengths through optical materials is helpful as are absorption qualities. Coupling of wavelengths with target materials is important, especially in materials (metals) processing and surgical applications. The selection of protective eyewear is dependent on the transmission and absorption, both, of the various wavelengths.

The wavelength chart, shown in Figure 3.11, shows three distinct regions of nonionizing radiation emitted by lasers. The central portion, called the "ocular focus" region, consists of wavelengths between 0.35 and 1.4 μm. All wavelengths

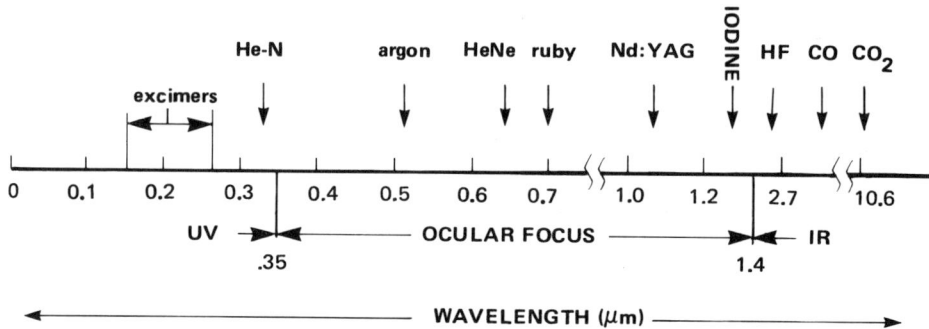

Figure 3.11 Wavelength chart.

in this range are focused by ocular components onto the retina of the eye by a magnifying factor of 100,000 times, or 10^5. In passing through the cornea, lens, and aqueous humor of the eye, essentially none of the beam of the laser is absorbed. Some question has arisen in research currently being conducted in the lower end of this range, near 0.35 μm, concerning an accumulation of retinal deterioration, so that long time exposures to the ultraviolet and even the blue light into the 0.55 μm region are discouraged without eye protection. Focusing of beams in this region onto the retina takes place at different locations, the shorter wavelengths focusing at shorter focal lengths than the longer wavelengths. This property of laser beam wavelengths is used in ophthalmology in determining which laser to employ, the argon type at 0.51 μm, or the ruby type at 0.69 μm, because the longer wavelength penetrates the retina deeper and can be used to "spot weld" detached retinas, whereas the shorter wavelength can cauterize bleeding capillary blood vessels near the surface of the retina.

Wavelengths outside the ocular focus region are absorbed by nearly all materials with few exceptions. It is particularly true as the wavelength extends from the extremities of this region of the spectrum. For example, the CO_2 laser wavelength is absorbed so well by so many optical materials that special single crystalline compounds (NaCl is one) must be prepared to transmit a high percentage of the beam.

CONTINUOUS BEAM (CW) PROPERTIES

A constant beam laser, known as a "continuous wave" or "constant wattage" (CW) laser is differentiated, of course, from a pulsed laser that provides bursts of energy. A CW laser beam intensity is measured in terms of the watts of power radiated, or contained. The beam undergoes little or no fluctuation with time creating a steady flow of coherent photons in a "CW mode." Many gas lasers can provide CW outputs because the gas can be engineered to flow through the lasing chamber not only to replenish the gas but also to provide cooling resulting from the excitation method, usually xenon lamps. CW lasers often require long periods of time (several minutes, or even hours) to stabilize the beam's intensity for very constant output. Essentially all alignment lasers are CW of low intensity emitting in the visual color range to be easily sighted.

In determining whether a laser beam is CW or pulsed, the American National Standards Institute, in the ANSI Z136.1 "Safe Use of Lasers" standard, defines a continuous wave as "The output of a laser which is operated in a continuous rather than a pulsed mode. In this standard, a laser operating with a continuous output for a period $\geqslant 0.25$ s is regarded as a CW laser."

The intensity of a CW laser beam is measured in the unit, watts per square centimeter (W/cm^2). This unit is often referred to as "power density" rather than the technical term, irradiance, used by physicists and defined in ANSI Z136.1 as the "Quotient of the radiant flux incident on an element of the surface containing the point at which irradiance is measured, (divided) by the area of that element. Unit: watt(s) per square centimeter, $W \cdot cm^{-2}$." In this work, the term power density and the unit W/cm^2 will be used to describe the intensity of a CW beam.

PULSED LASER BEAM TRAITS

The definition of a pulsed laser in this book will be that used in ANSI Z136.1, "A laser that delivers its energy in the form of a single pulse or a train of pulses. In this standard, the duration of a pulse (is) <0.25 s." The term "energy density" will be used to describe the intensity of a pulsed laser beam in terms of joules per square centimeter (J/cm^2). The ANSI Z136.1 standard defines a pulsed beam's intensity as "radiant exposure" and states it thusly: "Surface density of the radiant energy received. Unit: joules per centimeter squared ($J \cdot cm^{-2}$)".

The relationship of joules (energy) to watts (power) is:

J = W × time (one second), one watt-second,   or

$$W = \frac{J}{\text{time (one second)}}, \text{ one joule per second;}$$

thus fast-pulsed laser users can claim that the power (W) used by entire cities can be contained in one pulse of their laser. For example, in laser fusion research (inertial confinement experiments), if a laser is pulsed for one nanosecond, 10^{-9} sec, and contains 40 kJ energy, the power calculation would yield:

$$W = \frac{J}{s} = \frac{40,000 \text{ J}}{10^{-9} \text{ sec}} = 4 \times 10^4 \times 10^9$$

$$= 4 \times 10^{13} \text{ watts, or 40 Tarawatts!!}$$

To calculate the length of the light "bullet" for a 10^{-9} sec pulse, the simple calculation would be

$$= \frac{\text{speed of light (m/sec)}}{\text{sec}}$$

$$= \frac{3 \times 10^8 \text{ m/sec}}{10^{-9} \text{ sec}} = 3 \times 10^8 \times 10^{-9}$$

$$= 3 \times 10^{-1} = 0.3 \text{ m (or } \sim 1 \text{ ft)},$$

so that all the energy of that pulse would be contained in one foot of the laser radiation. Now the intensity of the beam in terms of J/cm^2 would have to be determined by the area of the surface receiving the energy. If all that energy (40 kJ) could be focused on a 100 μm spot on a target, that spot would receive

$$\frac{40 \text{ kJ}}{\text{spot area}} = \frac{4.0 \times 10^4 \text{ J}}{\pi(100 \times 10^{-6})^2/4 \text{ cm}^2}$$

$$= \frac{4.0 \times 10^4 \text{ J}}{.8(10^{-4})^2 = .8 \times 10^{-8} \text{ cm}^2}$$

$$= \frac{4.0 \times 10^4 \text{ J}}{.8 \times 10^{-8} \text{ cm}^2} = 5 \times 10^{12} \text{ J/cm}^2$$

$$= 5 \text{ terajoules/cm}^2 !!$$

Duration of the pulse, or "pulse width," is technically defined as the total time required for the pulse to rise from zero intensity, build to a maximum, and then fall to zero intensity again. Several methods of obtaining pulses of laser light are available, and control of pulse widths from a few milliseconds to femtoseconds is possible. Because these techniques can be quite complicated, especially in providing very short pulse widths, only a cursory description of some of the procedures will be presented.

Normal Pulsed Mode

This method provides a pulse by using stored energy to pump the lasing medium to an excited state temporarily, during a quick discharge of electrical energy, for example. A single burst of coherent light will emerge for the time excitation caused stimulated emission. Time durations obtained are generally in the millisecond domain for these normal pulsed modes.

Rotating Prism

This technique, shown diagrammatically in Figure 3.12 involves replacing the feedback mechanism or back reflector in the lasing chamber with a high reflectance

PULSED LASER BEAM TRAITS

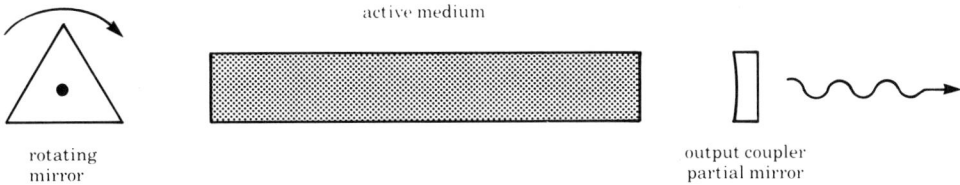

Figure 3.12 Rotating mirror pulsing technique.

mirror mounted on a rotating mount. In operation, the coherent light is reflected back and forth in the active medium until the oscillations build up sufficiently to pass through the output coupler (partial mirror). Because of the mechanical limitation of the rotating mirror, this method of pulsed laser light is limited to pulse widths in the microsecond region of time.

Q-Switching

This is actually a misnomer, or at least misleading, if the technical details of the lasing medium is not understood during the excitation period. The idea is to withhold stimulated emission until sufficient excitation takes place to "switch" a low Q factor to a high Q factor, and when the system in the cavity "relaxes," a burst of energy is released in a short period of time. (The Q-factor stands for "quality factor," used in electronics theory terminology.) Times on the order of 30 nanoseconds (ns) are possible by this method.

Faster Q-switching can be obtained by using Q switches that depend on certain electro-optic effects, such as those that take place in a Pockels cell. The cell, containing one of several specialized dyes in conjunction with an electro-optic switch, can produce pulse widths of a few picoseconds (ps).

One method of pulsing CO_2 laser beams for the early inertial confinement experiments at Los Alamos is described in an article in the January 1978 issue of EOSD Magazine. "Beyond the output coupler of the oscillator lies a 'scraper plate' (a large aperture), which passes most of the beam but reflects the outer portion of the beam into a laser-triggered spark gap. This device triggers a three-stage CdTe Pockels cell array, which acts as an electro-optic switch. The beam path is folded in a Z configuration spaced so that the beam's time of flight from the output of one Pockels cell to the input of the next corresponds with the transit time of the next high voltage cable delivering the trigger from one Pockels cell to the next. In this manner, the whole switch-out system acts as a traveling wave switch, increasing the contrast ratio well above that attainable with a single switch, resulting in a 1 ns pulse."

Ultrafast Pulsing

There are various electro-optical and special cell methods to provide ultrafast pulses of laser light that currently are approaching femtoseconds, 10^{-12} sec. References are provided for researching these highly technical processes.

Repetitively Pulsed Lasers

If a laser beam is repetitively pulsed, the ANSI Z136.1 Standard requires the evaluation of the laser beam's hazard as both a continuous beam and as a pulsed beam. The repetition rate, in pulses per second, pps, is measured in units called hertz which, in the case of one pulse per second, would be 1 Hz.

BEAM CROSS-SECTIONS, OR "MODES"

As mentioned previously, oscillations are intensified in lasers by multiple passes of coherent electromagnetic waves moving back and forth through the active medium. This is usually achieved by placing the active medium between two mirrors which act as a feedback mechanism. A number of different combinations of mirrors (plane, curved) have been used in practical lasers. The pair of mirrors, axially arranged about an intervening volume, is sometimes called an "optical cavity" or "laser resonator," because only certain frequencies of radiation will set up standing waves within it. These allowed frequencies of oscillation are called the axial, or "longitudinal modes," of the cavity. The amplitude distribution of the electric field in a plane perpendicular to the axis of the optical cavity is described as the "transverse electromagnetic (TEM) modes," which are actually cross sections of the beam.

A laser, as described earlier, is essentially an amplifying medium placed between two mirrors, as shown in Figure 3.13. The presence and shape of the mirrors (plane, spherical, etc.) fix the spatial distribution of the fields inside the cavity.

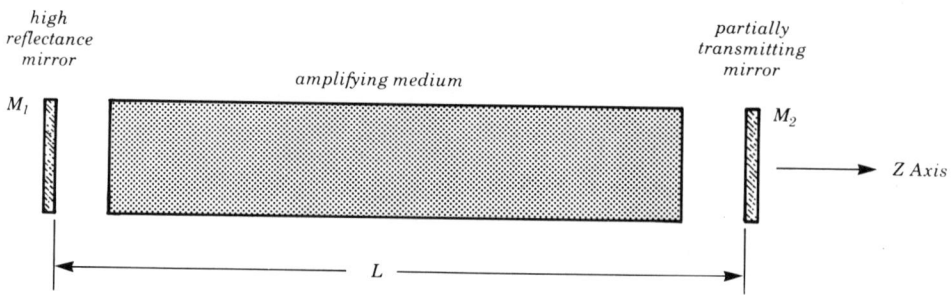

Figure 3.13 Optical resonator.

BEAM CROSS-SECTIONS, OR "MODES"

The light waves which are reflected back and forth inside a laser cavity produce what is known as a "standing wave pattern." Consider what happens when a laser beam of given amplitude and frequency encounters a mirror. The surface of the mirror acts as a boundary between two media. At this boundary the wave undergoes a reflection, sending a wave in the opposite direction. Interference may occur between the incident wave and the reflected wave if a certain condition is satisfied. If interference takes place, the resultant wave pattern appears to stand still. When two traveling waves of the same amplitude and frequency, moving in opposite directions interfere with one another, standing waves are produced. An example is given in Figure 3.14. Maximum values of the wave amplitude, called "antinodes," are shown, as well as positions of zero amplitude, called "nodes."

The situation described above is completely analogous to that of a vibrating string whose ends are tied down, as, for example, a vibrating guitar string. This, too, is an example of a resonant cavity. If the ends of the string are fixed in position, those standing waves which do oscillate must have nodes at the end points. This implies that there is a restriction on the wavelengths (frequencies) of those waves which will set up standing waves in the string.

Standing waves will be set up between mirrors in an optical cavity only if a certain relationship is satisfied between the wavelength of the light and the length L of the cavity. Given the length of a lasing cavity, one can determine the allowed frequencies of oscillation. Waves satisfying calculated allowed frequencies or "modes of oscillation" and directed along the axis of the laser are axial, or longitudinal, modes.

The longitudinal modes indicate how the electric and magnetic fields constituting the beam inside the cavity vary with the Z coordinate, in general, with the spatial coordinate located along the direction of propagation of the wave. However, the E-M field will be a function of the three spatial coordinates, X, Y, and Z, in general.

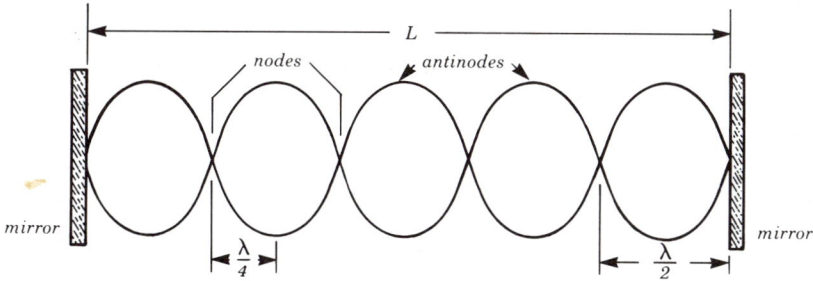

Figure 3.14 Standing waves in optical cavity.

The modes of oscillation which describe the functional dependence of the E-M field on X and Y coordinates are the "transverse modes." In a transverse mode, the electric and magnetic fields are each perpendicular to the direction of travel of the wave. Often in the literature, one sees the notation TEM_{lmn} modes with the subscripts l, m, and n. Physically, the subscripts l, m, n specify the *number* of *times* the electric (or magnetic) field crosses the x, y, and z axes, respectively. That is, the subscript indicates the number of nodes in the standing wave pattern along each coordinant axis. The subscript n is often dropped because the number of longitudinal modes is usually so great compared with the transverse mode numbers (e.g., $n \sim 10^6$, $l=1$, m=0). The notation is then abbrevated: TEM_{lm}.

To provide an insight into the physical meaning of TEM_{lm} modes, several patterns are presented in Figure 3.15.

These patterns are possible "beam profiles" or "cross-sectional" views. The transverse mode which is most commonly observed in general is the TEM_{00} mode. This pattern is the familiar "red spot" cross section of a low-powered He-Ne laser. There are no nodes in either the X or Y directions. The intensity of the laser beam does not fall to a minimum in either coordinate direction. The TEM_{10} mode shows that the intensity goes through a minimum (falls to zero) once in the X direction, and goes through no minimum in the Y direction. When various TEM_{lm} modes are observed experimentally, notice that the regions where the intensity pattern goes through a node (minimum) will be dark, whereas the bright patches indicate antinodes (maximum), or regions where beam power is concentrated. Also note that each time the pattern intensity goes through a minimum, the electric field has undergone a phase shift of π radians (180°). The TEM_{21} mode indicates that there are two minimums in the X direction and one in the Y direction.

The TEM_{01} or "doughnut" mode can be thought of as a linear superposition of the TEM_{01} and TEM_{10} modes. In practice, it is often obtained when there is a small particle of dust on one of the laser mirrors.

If you have trouble visualizing the various TEM modes and their physical interpretation, consider the following analogy. Suppose we have a stretched flexible membrane, a drumhead, for example. When the drumhead is struck at a given point, a wave travels outward from the point struck to the rim (boundary) of the drumhead where it undergoes a reflection. Waves are reflected back and forth in the drumhead. Under certain conditions (wavelength restrictions), interference may occur between incident and reflected waves, producing a standing wave pattern. This is similar to the case of a vibrating string, except that two-dimensional pulses are produced and the wavelength condition for allowed modes is not simple to state mathematically.

Optical cavities are usually designed to favor oscillation in *one* transverse mode, namely, TEM_{00}. The TEM_{00} mode is also called the "uniphase mode," the "fundamental mode," and the "lowest-order mode" (mode with the smallest

BEAM CROSS-SECTIONS, OR "MODES"

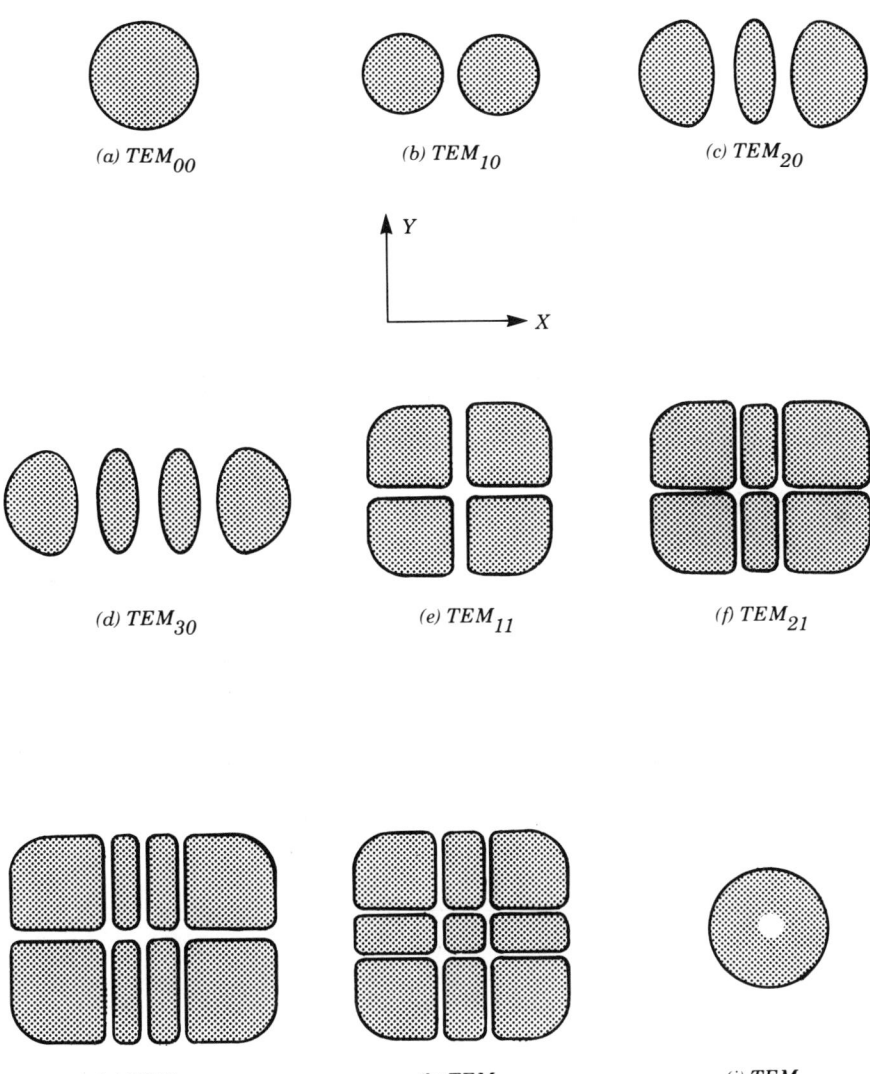

Figure 3.15 Selected TEM$_{lm}$ mode patterns.

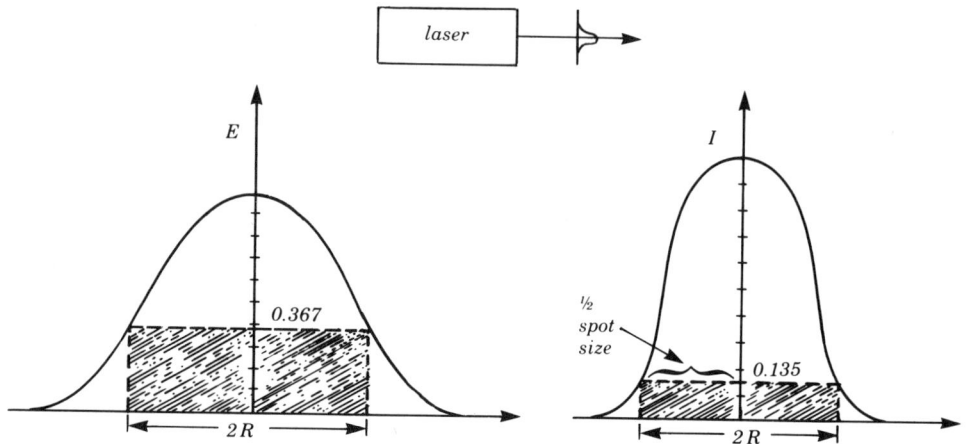

Figure 3.16 Electric field and intensity distributions in the TEM_{00} mode.

values of l and m). The cavity is said to be "lossy" for higher-order TEM_{lm} (modes with $l>0$, $m>0$).

To understand the importance of the TEM_{00} mode, one must observe the electric field and intensity distributions for a laser operated in this mode. Examine Figure 3.16. The bell-shaped curves, called Gaussion distributions of electric field, E (V/m), and intensity, I (W/m^2), are each plotted as a function of radial distance, R, *across* the beam.

There are several reasons for the desireability of the uniphase mode (TEM_{00}). One has to do with applications of lasers in materials processing (drilling, cutting, welding, etc.). The power density (power/unit area) that one is able to concentrate at the surface to be worked is a prime consideration. It can be shown that the beam from a laser operated in the TEM_{00} mode can be focused with an optical system down to a "diffraction-limited" spot size. This results in the maximum power density. The inability to focus high-order transverse modes to a diffraction-limited spot is due to the phase changes across the beam in such modes, and resultant "spreading out" of the beam power.

BEAM DIVERGENCE

The light emitted by a laser is confined to a rather narrow cone. As the beam moves through space, it slowly diverges or fans out. Figure 3.17 shows the way a circular beam diverges. At the output aperture of the laser, the beam diameter

BEAM DIVERGENCE

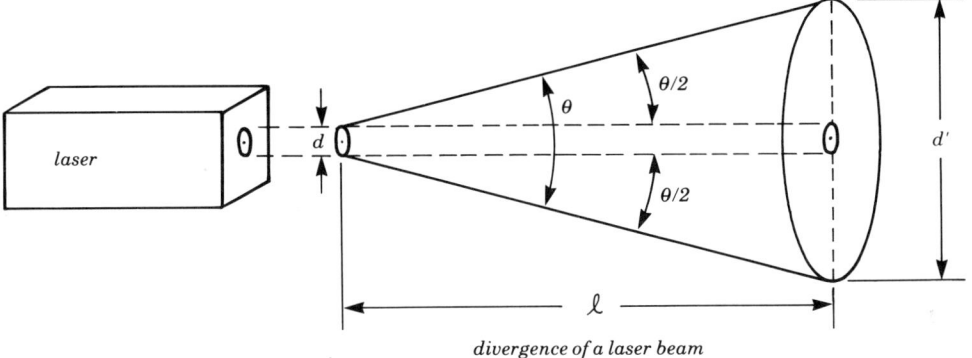

divergence of a laser beam

Figure 3.17 Divergence of a laser beam.

is d. Its beam divergence angle is θ, usually expressed in milliradians.* In traversing a distance ℓ, the beam diverges to a circle of diameter d'. If θ is a small angle, the diameter d' of the beam at a distance from the output aperture is given by:

$$d' \cong \ell\theta + d$$

For example, if ℓ=10 meters (10^3 cm), d=1 mm, and θ=1 milliradian, then

$$d' = (10^3 \text{ cm})(10^{-3} \text{ rad}) + 0.10 \text{ cm} \quad \text{and}$$

$$d' = 1.10 \text{ cm}$$

In manufacturer's specifications for a laser, the "half-angle beam divergence" ($\theta/2$ see Fig. 4.15) is often given as the divergent angle of the output beam. When using lasers, to avoid confusion, one should check carefully to see what is meant in a given case by the term "beam divergence."

The beam divergence angle varies considerably from one family of lasers to another. CW gas lasers generally have the smallest values of beam divergence (1 mrad or less). Solid-state and organic dye lasers usually have larger beam divergence angles (5-20 mrad). Semiconductor lasers emit a relatively wide cone of radiation (as much as 30°, or 0.524 radian, of full-angle beam divergence).

BEAM FOCUSING

Some of the most important uses of lasers depend upon their unique focusing properties. Some lasers are capable of being focused to produce what is called

---

*2π radians = 360°; 1 radian $\cong$ 57.3°. Hence 1 milliradian $\cong$ 0.0573° or 3.44 minutes of arc.

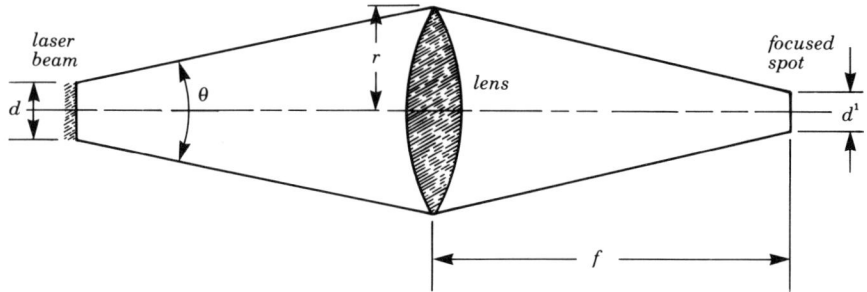

focusing of light with a simple lens

Figure 3.18 Focusing of a laser beam with a simple lens.

a "diffraction-limited" beam spot diameter. This is the smallest spot size attainable by means of a given optical system. Figure 3.18 displays the focusing of a laser beam by means of a simple lens of radius, r, and focal length, f. The beam has full-width angular divergence, θ. In geometrical optics, it is shown that the diameter, d', of the focused beam is approximately equal to the focal length of the lens, f, multiplied by the divergent angle, θ, or

$$d' \cong f\theta.$$

In Figure 3.18, d is the effective diameter of the laser beam at its output aperture.

To compute the amount of power per unit area concentrated on the spot using a laser as the light source, we first calculate the area A_{spot} of the focused beam:

$$A_{spot} = \pi \frac{d'^2}{4}$$

Substituting $f\theta$ for d',

$$A_{spot} = \frac{\pi(f\theta)^2}{4} = 0.785(f\theta)^2$$

A pulsed laser of peak output power, P_{max} (watts), has an average irradiance, or power density, E (W/cm²), at the focused spot of:

$$E = \frac{P_{max}}{A_{spot}} = \frac{P_{max}}{0.785(f\theta)^2}$$

The power density is inversely proportional to the square of both the focal length of the lens, f, and the angular beam divergence, θ. As value of the divergence angle increases, a lens of short focal length will concentrate more power on the focal area, A_{spot}. For example, consider a Q-switched Nd:YAG laser of peak power

BEAM FOCUSING

100 megawatts (MW) with angular divergence of 10 mrad. If a lens of focal length, f, of 25 mm is employed to focus the beam from this laser, the average power density at the focused spot would be:

$$E = \frac{100 \times 10^6 \text{ W}}{0.785 \, (2.5 \text{ cm})^2 \, (10^{-2} \text{ rad})^2} = \frac{10^8 \text{ W}}{0.785 \, (6.25 \text{ cm}^2)(10^{-4})}$$

$$= \frac{10^8 \text{ W}}{4.91 \times 10^{-4} \text{ cm}^2} = 2.04 \times 10^{11} \text{ W/cm}^2 \quad \text{or} \quad 204 \text{ gigawatts (GW)}$$

Can a laser beam continue to be focused indefinitely, to get smaller and smaller beam spot diamters? No, there is a limit to which a laser beam can be focused. This is the "diffraction-limited" beam spot size.

The intensity pattern of a laser in the TEM_{00} mode propagates as a Gaussian distribution for a distance of the order of d^2/λ, where d is the diameter of the laser output mirror. This is called the "near-field" pattern. At distances greater than d^2/λ, the beam naturally diverges, giving rise to an intensity distribution less well defined, the so-called "far-field" pattern. The full width, diffraction-limited beam divergence angle, θ_{diff}, in the far field is given by the formula,

$$\theta_{diff} = 2.44 \frac{\lambda}{d}$$

When focusing the beam with a simple lens (Fig. 3.18), the diameter, d', of the focused spot depends upon the full-angle beam divergence, θ, as well as the focal length, f, of the lens, in this relationship:

$$d' = f\theta.$$

Substituting θ_{diff} for θ in the above equation, we obtain the diffraction-limited beam spot diameter, d', as follows:

$$d'_{diff} = 2.44 \frac{f\lambda}{d}$$

The diameter of such a spot contains approximately 80% of the beam energy for a laser in the TEM_{00} mode. when the focal length, f, equals the beam diameter, d, then the diffraction-limited spot size, d'_{diff}, can be represented as:

$$d'_{diff} = 2.44\lambda$$

The focal spot area, A_{spot}, in this instance would be

$$A_{spot} = 0.785(d'_{diff})^2$$
$$= 0.785(2.44\lambda)^2$$
$$= 0.785(5.95)\lambda^2$$
$$= 4.67\lambda^2$$

Therefore, the beam radiated by a laser equipped with diffraction-limited optics can be focused down to a spot-size diameter approximately equal to 5 times the square of the laser's output wavelength.

Example: What is the diffraction-limited beam divergence of a Q-switched Nd:glass laser (λ=1.06 μm) having an output aperture d=1"? The calculation would be:

$$\theta_{diff} = 2.44 \frac{\lambda}{d}$$

$$= 2.44 \frac{(1.06 \times 10^{-6} \text{ meter})}{(2.54 \times 10^{-2} \text{ meter})}$$

$$= 1.02 \times 10^{-4} \text{ radian}$$

$$= 0.102 \text{ mrad}$$

What is the diffraction-limited spot-size diameter if the beam is focused by a lens of focal length, f, of 1.5"?

$$d'_{diff} = 2.44 \frac{f\lambda}{d}$$

$$= 2.44 \frac{(38 \text{ mm})(1.06 \times 10^{-3} \text{ mm})}{25.4 \text{ mm}}$$

$$= 3.87 \times 10^{-3} \text{ mm} = 4 \; \mu\text{m}$$

In calculating the diffraction-limited spot-size diameter of a 1-mm HeNe laser output beam, using the human eye focusing system, having a focal length of approximately 17 mm, the following equation can be written in determining d'_{diff} on the retina:

$$d'_{diff} = \frac{2.44 \; f\lambda}{d}$$

$$= \frac{2.44 \; (17 \text{ mm})(0.63 \times 10^{-3} \text{ mm})}{(1 \text{ mm})}$$

$$= 26 \times 10^{-3} \text{ mm} = 26 \; \mu\text{m}$$

This is a useful number in determining the intensity of beams of small diameters of wavelengths that are focusable by the human ocular system. If we calculate the magnification or intensification of a 5-mW HeNe laser beam of 1 mm diameter, the limited spot size of 26 μm limits the magnification factor as follows:

$$X(\text{magnification factor}) = \frac{\text{original diameter}}{\text{focused diameter*}}$$

---

*Diffraction-limited diameter focusable on the retina.

$$= \frac{1 \text{ mm}}{26 \times 10^{-3} \text{ mm}}$$

$$= 0.038 \times 10^3$$

$$= 38$$

Consequently, a 5-mW, CW, HeNe laser with a 1-mm diameter beam, can only focus 80% of the 5 mW on a 26-μm spot size on the retina. The intensity of radiation on the retina would be calculated as follows:

$$\text{Power density (W/cm}^2) = \frac{W}{A_{spot}}$$

$$= \frac{4 \times 10^{-3} \text{ W}}{.785 \, (26 \times 10^{-3} \text{ mm})^2}$$

$$= \frac{4 \times 10^{-3} \text{ W}}{.785 \, (26 \times 10^{-2} \text{ cm})^2}$$

$$= 70 \text{ mW/cm}^2$$

The damage threshold value for retinal tissue irradiation of the 0.63 μm wavelength is on the order of one watt per square centimeter, or 1000 mW/cm^2. Such small diameter HeNe laser beams, therefore, pose no potential for eye damage, therefore they can be employed in many applications without exposing the general public to possible eye hazards.

REFERENCES

1. DaMommio, A., D. M. Hull, et al. (1972). "Course 1, Introduction to Lasers, Laser Electro Optics Technology Series," Technical Education Research Centers, Inc., Cambridge, MA.
2. Jenkins, F. A. and E. H. White (1957, 1976). *Fundamentals of Optics*. McGraw-Hill, New York.

4
Measurement of Laser Beam Characteristics

Most applications of laser technology can employ a number of commercially available standardized laser systems that are ready for immediate use. These manufactured lasers have been characterized very carefully so that users need not be too concerned about actually making precise measurements of the various properties of the beam. However, a knowledge of the methods of measurement is helpful, so a very brief description of several instruments used in the field of laser beam measurements is presented. (See references for more detailed information about this specialized field of physics.)

WAVELENGTH MEASURING DEVICES

An optical instrument used to measure the wavelengths or frequencies of light emitted by various light sources is commonly known as a "spectrometer" or "spectrograph." In the former case (spectrometer), the emphasis is placed on using the instrument to *measure* the different wavelengths of the spectral lines, while in the latter case (spectrograph), the emphasis is placed on using the instrument to *record* or *photograph* the spectrum. If the instrument is simply used to *observe* the spectrum of a light source, it is often referred to as a "spectroscope." Regardless of the manner in which this particular optical instrument is used, its construction is essentially the same. All spectrometers depend on an important optical element, usually a prism or grating, to separate the light into its individual colors or wavelengths. Once separated, the different wavelengths can be recorded and measured.

Prism Spectrometer

An instrument composed of a collimator, prism table, and a telescope is called a "prism spectrometer." Operation of the spectrometer is described here. Light enters through a slit at one end of the collimator and emerges parallel to the axis of the collimator at the other end. The collimated light then strikes a prism which has been appropriately positioned on the spectrometer table. The prism disperses the collimated light into different wavelengths of light that are present in the light source. The sets of parallel rays (one set for each wavelength) then enter the telescope and form separate, distinct images of the collimator slit in the focal plane of the telescope. The separate slit images (spectral lines) are viewed by an observer looking through the telescope. Cross-hairs in the telescope may be positioned over each spectral line, thereby fixing the angular position of the telescope for each wavelength. The angular position of the telescope is determined with the aid of a circular scale which is graduated in fractions of a degree. Thus a direct relationship exists between the position of the telescope and a given wavelength of light passing through the prism. It is this relationship which permits the prism spectrometer to be used to determine the wavelength of light.

As a ray of monochromatic light passes through a prism, it is refracted twice, once as it enters and once as it leaves the prism, as shown in Figure 4.1.

The angle A between the two prism surfaces or planes where refraction takes place is called a "prism angle." The intersection of the two refracting planes is

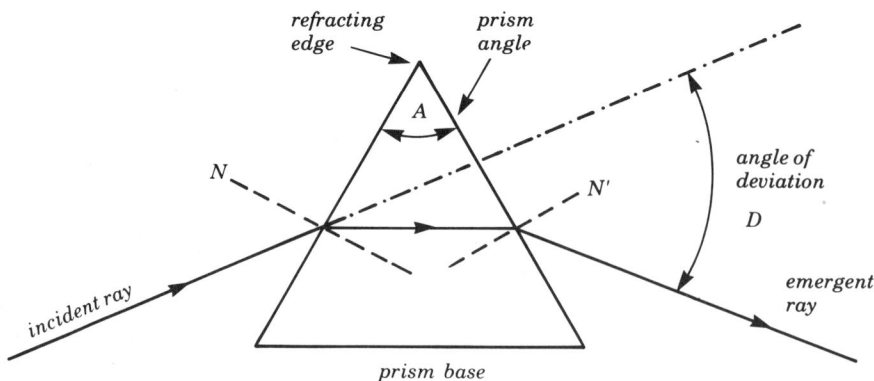

Figure 4.1 Refraction of light by a prism.

WAVELENGTH MEASURING DEVICES

called the "refracting edge" of the prism. The base of the prism is that side of the prism opposite the prism angle. (This name may be misleading; it is important to realize that the prism does not rest on the base during normal operation.) The normals to the prism faces are shown an N and N'. As the figure clearly shows, the refracted and incident rays are in different directions; the angle D between the incident and emergent ray is defined as the "angle of deviation."

If a beam of collimated light, containing several wavelengths, is incident on the prism face, different wavelengths will be refracted by different amounts and will therefore necessarily have different angles of deviation. The particular angle of deviation depends on the prism angle, the index of refraction of the prism at that wavelength, and the angle of incidence. The angle of deviation is a minimum if the angle of emergence is equal to the angle of incidence, a condition which can be easily achieved with a prism spectrometer. When the prism spectrometer is set at minimum deviation for a given wavelength, it can be shown that the index of refraction, n (at that wavelength) is given by:

$$n_\lambda = \frac{\sin \frac{A+D_{min}}{2}}{\sin \frac{A}{2}}$$

where A is the prism angle, D_{min} is the measured angle of minimum deviation for the particular wavelength, and n_λ is the index of refraction of the prism at that wavelength. By determining the angle of minimum deviation for each wavelength (slit-image in the focal plane of the telescope), one can calculate the appropriate n_λ (for the wavelength λ) from the equation just given. The graph of index of refraction versus wavelength is called a "dispersion curve." Several dispersion curves for various glasses are shown in Figure 4.2.

Grating Spectrometer

A "grating spectrometer" is similar to a prism spectrometer in every way with the exception that a simple plane diffraction grating replaces the prism as the dispersing element. The function of the collimator and telescope, as well as the general operation of the spectrometer, remains unchanged.

The essential makeup of a grating spectrometer is shown in Figure 4.3. The similarity between this arrangement and that of the prism spectrometer is readily apparent.

The diffraction grating is a simple and most useful instrument for studying spectra. The grating consists of a grid of fine parallel lines uniformly spaced on a polished reflecting or transmitting surface. The lines are generally ruled with a fine diamond point and generally number several thousand lines per centimeter. The reflection or transmission of light, by the uniformly separated portions of

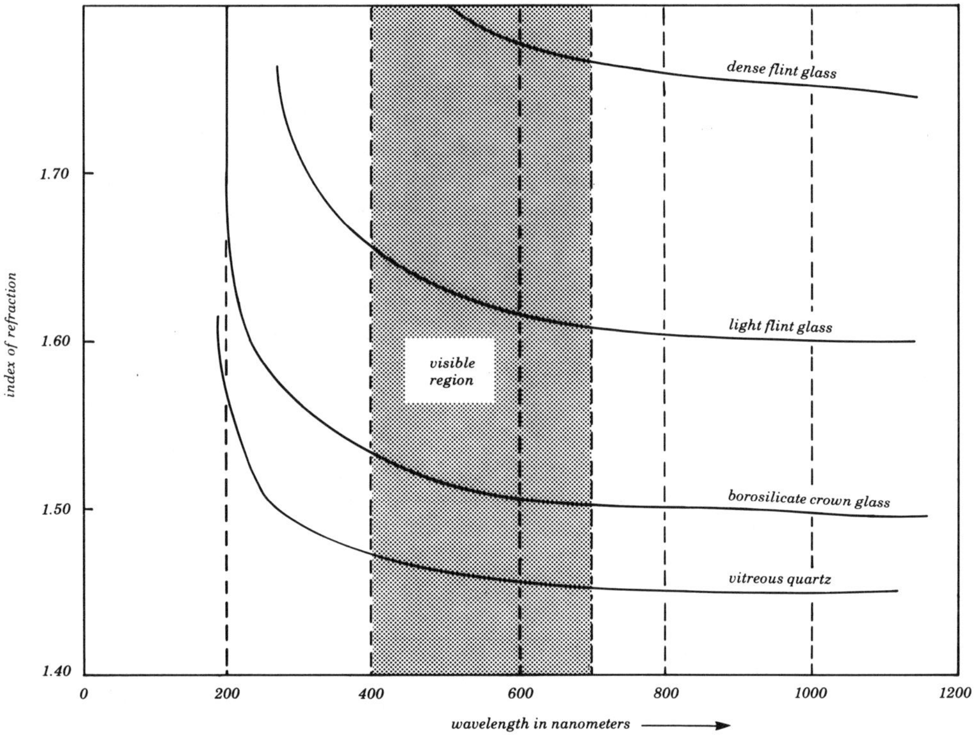

Figure 4.2 Dispersion curves for several different optical materials.

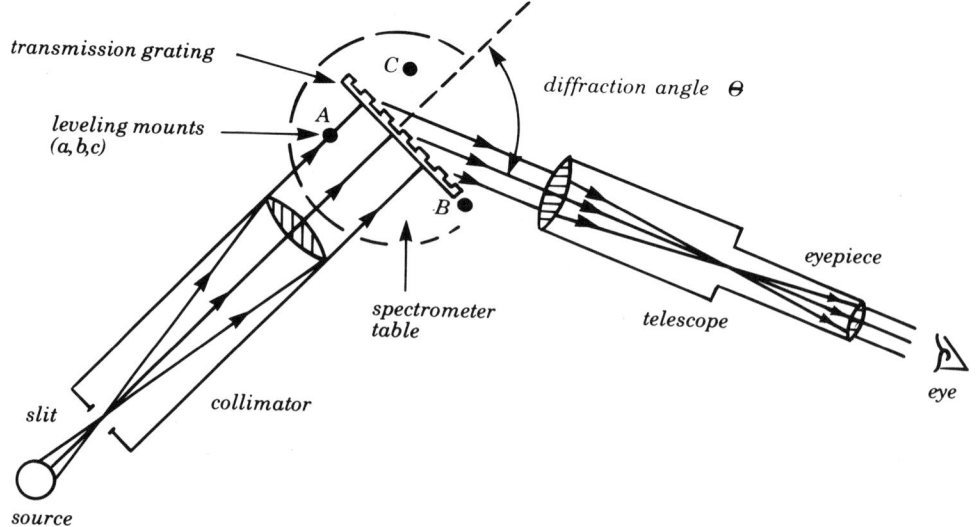

Figure 4.3 Grating spectrometer.

WAVELENGTH MEASURING DEVICES

the surface, causes diffraction and interference effects. These effects lead to the formation of a spectrum characteristic of a given light source.

The collimator directs parallel light onto the grating. The grating, in turn, serving as the dispersing element, separates the bundles of parallel light of different wavelengths and directs them into the telescope. The observer adjusts the telescope for proper focus and views the characteristic line or band spectrum of the source in the focal plane of the telescope.

The diffracting angle, θ, is unique for each wavelength of light in the light source. The wavelength of the light, λ, the diffracting angle, θ, the grating constant, d, and the order number, m, are all related in the grating equation:

$$m\lambda = d(\sin i + \sin \theta)$$

Here i is the angle which the incident collimated beam makes with a normal to the grating surface. The collimated beam is perpendicular to the plane of the grating so that for this orientation angle i is zero and sin i is zero. Thus for normal incidence, the grating equation reduces simply to:

$$m\lambda = d \sin \theta,$$

where m = ±1, ±2, ... refers to the different orders of diffraction. The essential geometry relating λ, θ, and d for order, m, is illustrated in Figure 4.4. Thus, for a given order other than the zero order (m=0), the incident light is diffracted at an angle, θ, given by $\sin \theta = m\lambda/d$. Since d is the grating constant and m is constant for a given order, the angle, θ, changes with wavelength. For each order (m=1, 2, 3, etc.), the different wavelengths are separated out and observed as a distinct

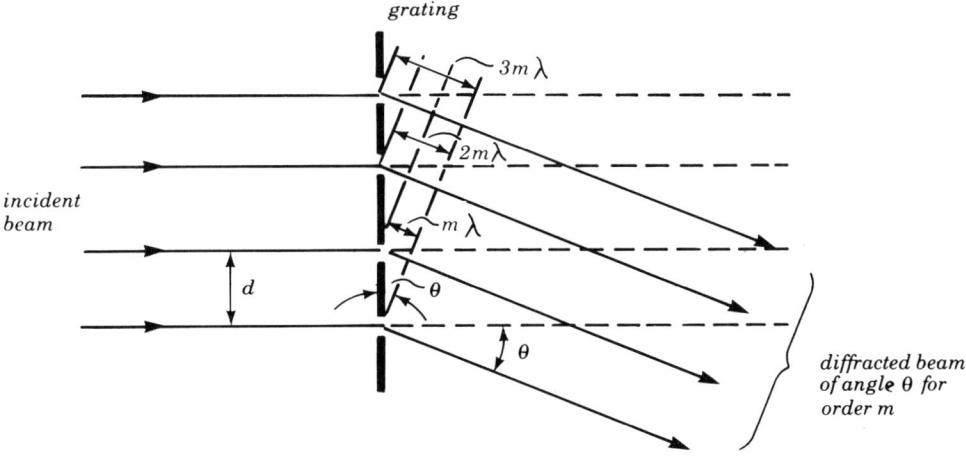

Figure 4.4 Diffraction for order, m, at angle, θ, at normal incidence.

spectrum. However, since spectral lines in adjoining orders may overlap, one must be careful in interpreting the spectral lines observed in the focal plane of the telescope. For example, a red line of 700 nm wavelength in third order is diffracted through the same angle as the green line of 525 nm wavelength in fourth order since $\sin \theta = 3(700)/d = 4(525)/d$ is identical for each wavelength. For visible light, there is no overlapping of the first and second orders since with $\lambda_1 = 720$ nm and $\lambda_2 = 400$ nm, the red end of the first order falls just short of the violet end of the second order. When photographic observations are made, however, these orders may extend down to 200 nm in the ultraviolet, and then the first two orders do overlap. One can avoid this complication through the use of suitable color filters to absorb from the incident light those wavelengths which would overlap the region under examination. For example, a cut-off filter which transmits only wavelengths longer than 600 nm might be used to block the shorter wavelengths at order $m=2$ or higher from disturbing the observations in the wavelength region near 700 nm. Attention to color, where orders do overlap, will help in avoiding incorrect identification of wavelengths in unknown spectra. However, in many cases, use is made of the overlapping order lines as wavelength standards for more exact assignment of unknown lines.

Monochromator

In many phases of optical work, it is necessary to obtain the transmittance, reflectance, or absorbance of solids, liquids, and gases. Occasionally, it may be required to determine the emittance of heat-radiating materials as well. Furthermore, it is desirable to obtain the variation in such quantities as a function of the wavelength. Such information allows the engineer to choose proper filters, antireflection coatings, thin film protective coatings for mirrors, etc., or, conversely, to determine the basic optical properties of dyes and gases which may be used in lasers. Measurement of emitted radiation can allow one to determine radiating species and their temperatures and concentration (density). Thus, it is necessary to have available an instrument which will provide a calibrated source of light at various wavelengths or act as a variable and calibrated filter. Such a device is called a "monochromator."

The basic principles used in the spectrometer are also used in the monochromator. In the spectrometer, the various wavelengths emerge at different angles and a telescope is rotated to observe each wavelength. In the monochromator, on the other hand, internal adjustments allow one to obtain all wavelengths successively at one exit slit. Thus, the monochromator could be thought of as a "black box" with a wavelength-adjusting knob and an input and output opening. The light source is placed at the input, the various wavelengths are dispersed in the box, and the wavelength chosen by the adjusting knob emerges at the output. The dispersion of the light can be accomplished in several different ways, for

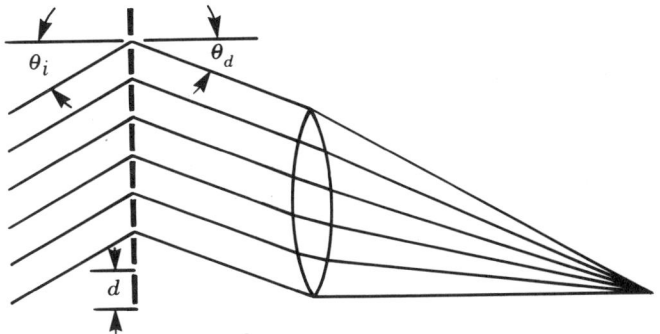

Figure 4.5 Transmission grating of a monochromator.

example, by a prism, a reflection grating, or a transmission grating. Consider a monochromator design based upon a transmission grating.

A beam of parallel light of wavelength, λ, is incident upon a transmission grating of grating constant "d" at an angle, θ_i, as shown in Figure 4.5. After diffraction, the light will emerge at some angle, θ_d, depending upon the wavelength of the light and the angle of incidence, θ_i. The mathematical relationship between θ_i, θ_d, d, and the wavelength, λ is given by:

$$m\lambda = d(\sin \theta_i + \sin \theta_d)$$

where m is the order of diffraction. This equation indicates various wavelengths can be successively obtained at a fixed exit slit, that is, a fixed angle, θ_d, by simply varying the incident angle, θ_i.

The basic design of a monochromator is shown in Figure 4.6. Light from a source is focused onto an entrance slit, S_1. Light from the slit, S_1, is reflected by the mirror, m, through the lens, l_1, which renders parallel, through the grating, D, where it is dispersed and then through the lens, l_2, which focuses it onto the exit slit, S_2. The lens, l_3, is present only as an auxiliary lens to collimate the light, after dispersion, into a beam. Variation of θ_i is produced by rotating the wavelength dial, which in turn rotates a cam, thereby tilting the mirror, m. In this way, various wavelengths can be obtained at the exit slit, S_2. This is the basic design used in the monochromator.

The pertinent characteristics of a monochromator are the wavelength range over which it can be utilized and the bandwidth of emerging light. The wavelength range is determined by the properties of the optical components and the angle over which the mirror can be rotated. Monochromators are commercially available which cover various wavelength ranges between 50 nm and several microns. The bandwidth is determined by the dispersion of the grating.

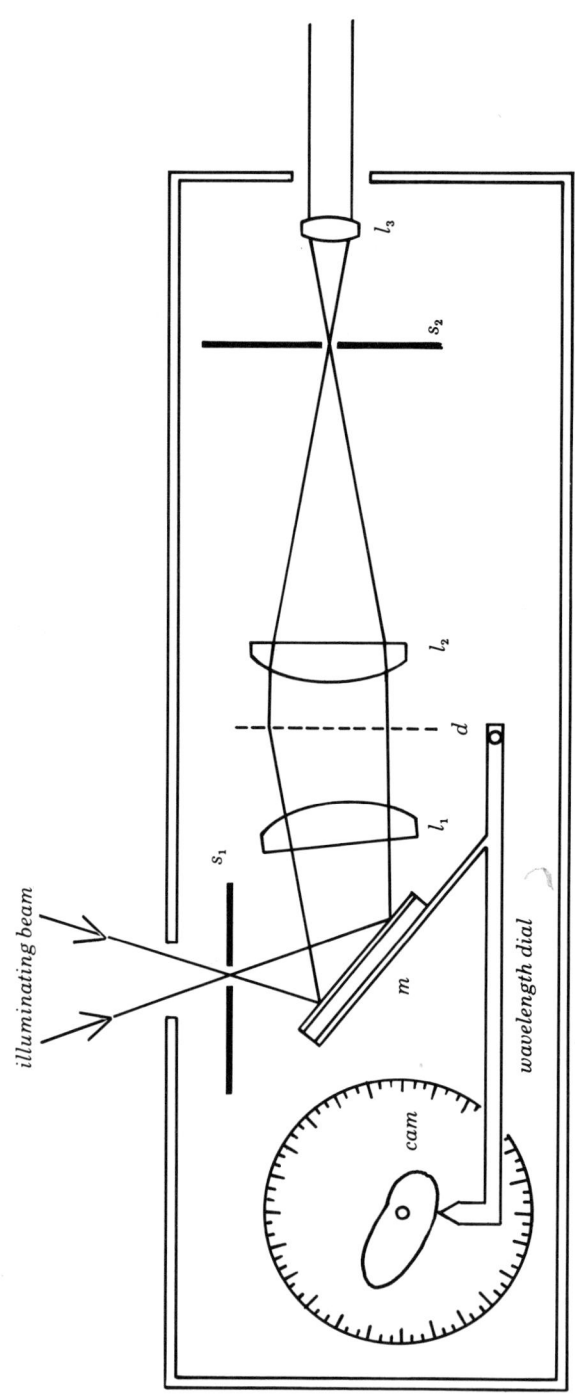

Figure 4.6 Basic design of a monochromator.

WAVELENGTH MEASURING DEVICES

A monochromator has two basic uses: (a) as a wavelength analyzer for an unknown source of light, or (b) as a wavelength source. A wavelength analyzer is required if one wishes to know the spectral distribution available from a laser light source. A monochromatic light source is required if one wishes to determine the transmittance, absorbance, or reflectance of a filter or antireflection coating as a function of wavelength.

When used as a wavelength analyzer, as shown in Figure 4.7, the light source is focused onto the entrance slit of the monochromator and the intensity of the light emerging from the monochromator is measured at each wavelength by a detector, which may include thermopiles, radiometers, photomultipliers, photodiodes, and phototubes. A problem which arises in the use of this type of instrument is that the detector is often not equally sensitive to all wavelengths. Thermopiles and photometers are preferred because they detect wavelengths with equal sensitivity and should be used for measurements where possible; however, their overall sensitivity is considerably less than that of a phototube.

When the monochromator is used as a monochromatic light source, the arrangement shown in Figure 4.8 is used. In this case, it would be desirable to have a source of light which emits a continuous spectrum with equal intensity at all wavelengths. Such a source is not available. As a result, one must either know the precise source spectrum or must provide for some mechanical or electrical method of correcting for the variation in source spectrum.

In measuring the "transmittance" of an optical filter, one utilizes the arrangement shown in Figure 4.9. In this measurement, several types of sources could be used. The output of the detector is amplified and measured by a meter.

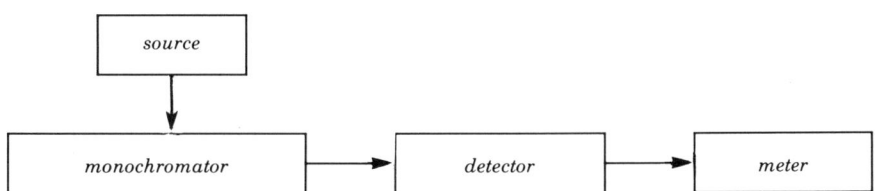

Figure 4.7 Block diagram for monochromator used as a wavelength analyzer.

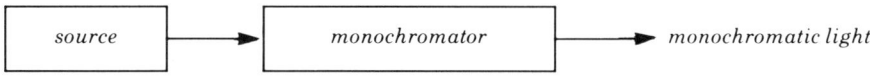

Figure 4.8 Block diagram for monochromator used as a monochromatic light source.

MEASUREMENT OF LASER BEAM CHARACTERISTICS

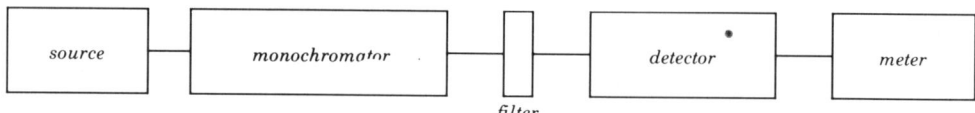
filter

Figure 4.9 Block diagram for monochromator used in measuring transmittance of an optical filter.

To correct for the source spectrum and detector response, the following procedure is used:

1. With no filter in the beam path, the desired wavelength on the monochromator is chosen and the amplifier gain is adjusted for a full scale reading
2. The filter is placed in the beam path and transmittance is read on the meter

This procedure is repeated for all wavelengths of interest. In this way the gain of the amplifier circuit is adjusted at each wavelength to overcome the variations introduced by the light source and the phototube. As a result, one can obtain a plot of transmittance as a function of wavelength which is a true representation of the filter.

Suppose the absorption spectrum of a dye suitable for use as a laser material is desired. The dye to be examined is Rhodamine 6G. The dye will be placed in a liquid solution and its absorption spectrum measured.

The equipment required to obtain an absorption spectrum consists primarily of:

1. A steady source of light for the wavelength region required
2. A means of reproducibly selecting appropriate small portions of this region
3. A mount or container which can be precisely aligned to hold the material under study
4. A detector which is sensitive over the wavelength region under consideration
5. A means of adjusting the intensity of radiation falling on the detector to provide a reference whereby intensities can be compared before and after traversing the specimen

In this experiment, these five requirements will be met as follows. The source of light will be a tungsten lamp; the wavelength selector, a monochromator; the detector, a phototube with amplifier and meter indicator. The intensity of the incident radiation can be adjusted by using the zero and reference controls. A

WAVELENGTH MEASURING DEVICES

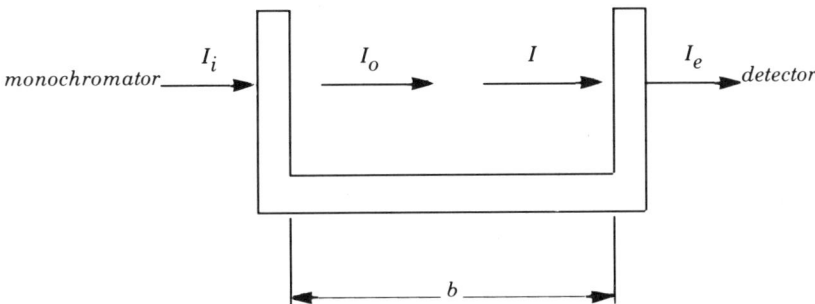

Figure 4.10 Arrangement for measuring the absorption spectrum of a laser dye.

sample compartment is added between the monochromator and detector. This arrangement is shown in Figure 4.10.

Light from the tungsten source is incident on the entrance slit of the monochromator and is dispersed by the grating. The particular wavelength of light is then chosen with the wavelength dial, passed through the dye sample in the sample compartment, and is detected by the phototube. The transmitted intensity is read on the meter. The sample chamber has provision for holding four 1-cm² absorption cells or "cuvettes." A selector knob on the top of the compartment permits placement of each of the cells in the light path. There is also provision for blocking the light path to the detector (by placing the selector knob at zero). This position is used to simulate zero transmittance.

To determine the absorption spectrum, one measures the incident intensity of the light passing through the cell containing the liquid. This is shown schematically in Figure 4.11, where I_i is the intensity of the light of wavelength which leaves the monochromator, I_o is the intensity of the light which enters the dye,

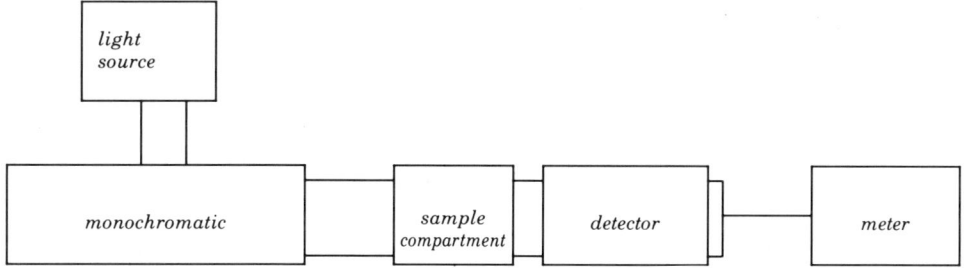

Figure 4.11 Schematic diagram used in calculation of absorption measurement.

I is the intensity of the light which leaves the dye, and I_e is the intensity of the light which enters the detector. I_o will be slightly less than I_i due to the fact that a small amount of light will be scattered and absorbed by the cell wall as the beam passes through it. I will be less than I_o due to the absorption by the molecules of dye in the cell itself. Passage of the light through the second cell wall will also be accompanied by some scattering and absorption, thus I_e will be less than I.

The absorption measurement is made in the following manner. The monochromator is set to pass a particular wavelength. The light path to the detector is blocked and the meter is set at 0.0 transmission with the zero adjust knob. An absorption cell containing the solvent which is used in making the dye solution is placed in the light path. The detector amplifier gain is adjusted with the reference control so that the meter reads 100% transmission. A second absorption cell, identical to the first but containing the dye solution, is placed in the beam path and the % transmission recorded. Since the 100% transmission point was set with cell plus solvent in the beam path, the measured % transmission of the second cell represents the transmission of the dye alone, in other words, the effect of absorption and scattering at the cell walls and in the solvent has been effectively cancelled out. Thus, the fraction of the light transmitted, T, (called the transmittance), is given by this equation:

$$T = \frac{I_e}{I_i} = \frac{I}{I_o}$$

and

$$\%T = \frac{I_e}{I_i} = 100\frac{I}{I_o}$$

In order to determine the %T as a function of wavelength, one must now repeat this procedure for each wavelength. In this way, a curve of %T versus wavelength can be obtained.

It is now necessary to relate the transmittance to the length of path through the dye and the dye concentration. Consider a dye cell of length b_1. In this length, the incident light intensity I_o will be reduced by some fraction, f, to I_1 or $I_1 = fI_o$. If the path length is doubled, the intensity I_1 will be further reduced by the fraction, f, to I_2 or $I_2 = f^2 I_o$. For n thicknesses, one thus has $I = I_n = fI_o$. If one then writes the equation

$$\frac{I}{I_o} = f^n$$

then

$$\log_{10}\frac{I}{I_o} = \log_{10} f^n = n \log_{10} f$$

where n equals the ratio of total path length, b, to incremental path length, b_1. Furthermore, since the transmittance is affected by the concentration of dye in the solvent, it is necessary to introduce the concentration into the equations. To do this, consider the following argument. Suppose one has a dye cell with length $b=2b_1$, thus $I=f^2 I_o$. The cell contains solvent with concentration, c, of dye in solution. If one could now reduce the thickness of the cell to one-half its original thickness while at the same time evaporating one-half of the solvent, we would have a cell of length, b_1, but with double the concentration, that is, 2c. Since no dye was removed, this half-cell of length, b_1, and concentration, 2c, must absorb in exactly the same way as the cell of length, $2b_1$, and concentration, c. Likewise, doubling the concentration while maintaining constant length will have the same effect as doubling the length and holding the concentration the same. Thus, if both concentration and length are included, $n=bc/b_1 c_1$, and

$$\log_{10} \frac{I}{I_o} = \frac{bc}{b_1 c_1} \log_{10} f$$

Here $(\log_{10} f)/b_1 c_1$ is a constant, the negative of which is defined as the molar absorptivity, a, that is

$$\text{Molar absorptivity, a} = -\frac{1}{b_1 c_1} (\log_{10} f)$$

Using the molar absorptivity, a, the equation becomes

$$\log_{10} \frac{I}{I_o} = bc(-a) \quad \text{and} \quad -\log_{10} \frac{I}{I_o} = abc$$

or

$$\log_{10} \frac{I_o}{I} = abc \quad \text{and} \quad \log_{10} \frac{1}{T} = abc$$

The quantity $\log_{10} (1/T)$ is defined as the absorbance, A. Thus, A=abc. So absorbance = (molar absorptivity) (cell thickness) (molar concentration). The relation, $\log_{10} I_o/I = abc$, is known as "Beers law."

Thus, by measuring the transmittance T of a substance, one can calculate the absorbance, A. Since b, the length of the cell is a known constant, one can, with the help of Beers law, calculate the absorptivity, a, if c is known, or the concentration, c, if a is known. Note that the molar absorptivity, a, is a constant independent of cell length and solution concentration. Values of the molar absorptivity can be found in chemical reference books. From these values, and a spectrophotometer measurement, the concentration of a solution is found. This is a standard procedure used in many laboratories.

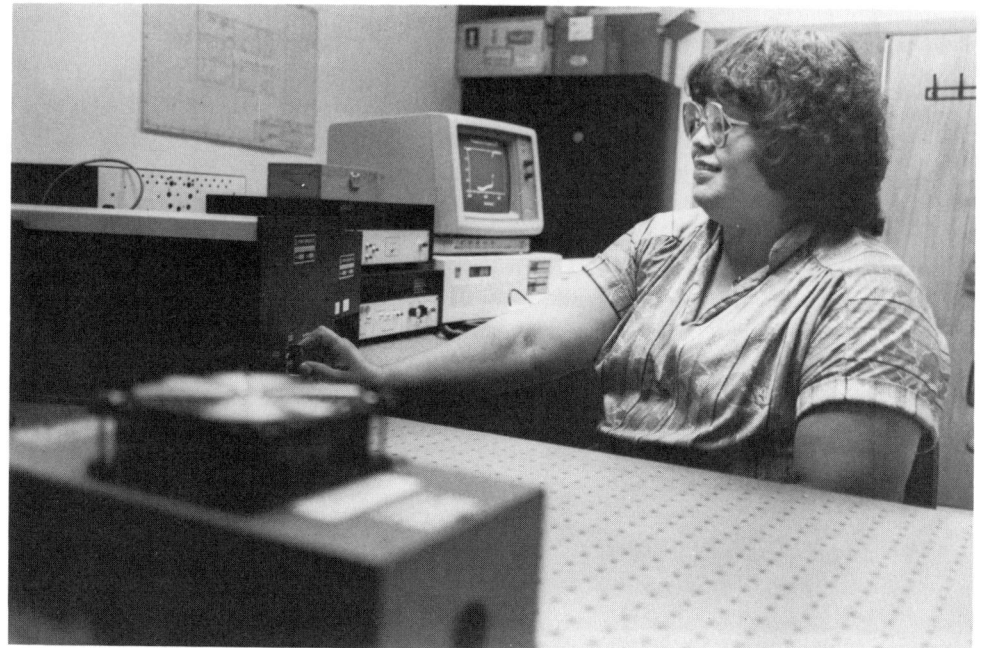

Figure 4.12 Photograph of technician obtaining an absorption spectrum curve for a plastic material using a spectrophotometer. (Courtesy of Newport Corp.)

Spectrophotometer

In passing through a translucent material, whether solid or liquid, a beam of light loses some of its original intensity because only part of the light gets through and part is absorbed. In absorption spectrophotometry one is concerned with the determination of the part of light intensity which is absorbed. A graph of the corrected light absorption as a function of wavelength is called an absorption spectrum of the material under study. Figure 4.12 shows a technician operating a spectrophotometer to determine the absorption spectrum of an orange plastic. Notice the curve on the video display unit. The light source is an argon laser.

POWER AND ENERGY MEASUREMENTS

A "photodetector" is defined as a device capable of providing an electrical or thermal response that is a useful measure of incident electromagnetic radiation.

An example of a visible radiation-type detector is one which produces an electrical current as a result of visible radiant energy (light) striking its surface.

Three basic components in any laser "photometric" or "radiometric" system are the laser, a sensor or detector (e.g., solar cell, photodiode), and display device (e.g., recorder, oscilloscope).

The optical source under observation may be transient, of short duration, a nonrecurrent event, or a repetitive action. One of the main considerations in building a system to record the source is the choice of a proper detector to meet certain requirements. In choosing an optical detector, one must take into consideration many factors which may influence the measurement. Manufacturers of photodetectors have, in general, followed international standards of measurement of describing these characteristics. Some of the more important characteristics are (a) spectral response, (b) responsivity, (c) noise, (d) response time, (e) linearity, and (f) quantum efficiency.

General interest in fast response detectors for use in the ultraviolet, visible, and infrared spectra has increased considerably since the invention of the laser and related electro-optical devices. Photoemissive devices, such as the vacuum phototube and photomultiplier, which for a long time have been used for observing short light pulses (e.g., scintillation counters) or fast intensity modulation (e.g., flying spot scanners), have more recently been used for detection of stimulated emission from lasers.

A "photoelectric effect" may occur when light strikes special combinations of materials. A voltage may be generated, a resistance change may take place, or electrons may be ejected from the material's surface. As long as the light is present, the condition continues; it ceases when the illumination is interrupted except for special cases, for example, "after pulsing" which occurs in photomultipliers.

Photodetectors can be divided into two major categories:

1. *Quantum detectors*, which respond directly to incident photons. These are uncharged packets or quanta of electromagnetic energy which is dependent on the frequency or wavelength of electromagnetic radiation. Quantum detectors may be of the following types:
   a. *Photoconductive* where the conductivity of the photosensor changes as a function of the incident radiation. There may be a linear or nonlinear relationship between incident energy and change in resistance. Under this category are found *bulk* photoconductors, such as photoresistors, which may be intrinsic or undoped, and *doped* photoconductors, such as photodiodes.
   b. *Photoemissive*, where incident photons free electrons from the surface of the detector. This phenomenon generally occurs in a vacuum photodiode, biplanar phototube, or photomultiplier.
   c. *Photovoltaic*, in which a voltage is self-generated as radiant energy strikes the surface of the device in which no external power source is

used. A good example of this type of device is the solar cell used on satellites and spacecraft to convert the sun's radiation into useful electrical power. The photodiode may also be operated in the photovoltaic mode by removing the reverse voltage bias.
2. *Thermal detectors* are those where the radiation absorbed is transformed into heat and the detector responds to the change in temperature. Examples of this type are the thermopile, thermistor, Golay cell, and bolometer. These are normally associated with the infrared spectrum.

Before discussing the theory of operation and application of photodetectors, it is important that we understand some of their limitations and how these limitations affect their response to different stimuli. Each type of detector may respond differently to the same stimulus, but, in general, all detectors have the following characteristics in common for a basis of comparison.

"Responsivity" may be defined as a measure of a detector's sensitivity to radiant energy and is given as the ratio of the signal current measured in amperes produced by the detector to the incident radiation measured in watts at the entrance to the detector. If the signal output is represented as a voltage, it may be given as the ratio of signal voltage per watt of incident radiation. Thus, the responsivity is essentially a measure of the effectiveness of the device to convert electromagnetic radiation to electrical current or voltage. The responsivity will vary with changes in wavelength, bias voltage, and temperature. Responsivity changes with wavelength since the reflection and absorption characteristics of the detector's sensitive material change with wavelength. Temperature changes affect both the optical constants of the detector material and its collection efficiency.

If we plot the absolute responsivity as a function of wavelength, we have a "spectral response" curve. This curve is commonly shown as a relative response curve in which the peak of the response curve (highest responsivity) is set equal to 100% and all other points are relative to the peak. A common practice is to make an absolute calibration only at the peak of the response curve. Thus, to determine the absolute responsivity at some other wavelength, we find the percent relative response at this wavelength and multiply it by the absolute value of amps/watt at the peak of the curve.

If a constant source of radiant energy is instantaneously turned on and irradiates a photodetector, it will take a finite time for current to appear at the output of the device and for the current to reach a steady-state value. If the same source is now instantaneously turned off, it will again take a finite time for the current to follow the change and decay back to its initial zero level. The term "response time" normally refers to the time it takes the photocurrent generated by the detector to rise to a value which is 63.2% of the final or steady-state value reached after a prolonged period of time. The recovery time is the time it takes

POWER AND ENERGY MEASUREMENTS

the photocurrent to fall from its steady-state value to a value which is 36.8% of the steady-state value when the detector was excited by radiant energy. Since photodetectors are apt to be used for detection of fast pulses, a more important term called "risetime" is used to describe the speed of response of the detector. Risetime is defined as the time difference between the 10% point and the 90% point of the peak amplitude output on the leading edge of the pulse. "Falltime" is measured between the 90% point and the 10% point on the trailing edge of the pulse waveform. This is sometimes referred to as the decay time. The time measured between the first appearance of current and the appearance of radiation incident on the surface of the detector is called the delay time. This time may be insignificant compared to the risetime. The sum of these two periods is called the "turn-on time." The time measured between the extinction of radiation incident on the detector and the first indication of change in the current output is the storage time. Adding the storage time to the falltime, we have what is called the turn-off time. The most significant of these pulse response times are the risetime and falltime, and these are the two terms most often quoted by manufacturers in the literature. The risetime of a device that transmits or displays waveforms is taken as the risetime of the output (or displayed) waveform if the device were driven with a theoretically perfect step function (zero risetime). A source whose risetime is much less than the risetime of the device under test is the accepted practice. A source with a risetime less than or equal to 1/10 of the risetime of the detector being tested would be ideal. Another factor to consider is the risetime limitation introduced by cables or of the display device, for example, the oscilloscope or recorder. Figure 4.13 shows a typical experimental setup for measuring the risetime of a fast-pulsed light source.

"Photodetectors" are characterized by a photocurrent response which is linear with incident radiation over a wide range. Any variation in responsivity with incident radiation represents a variation in the linearity of the detector.

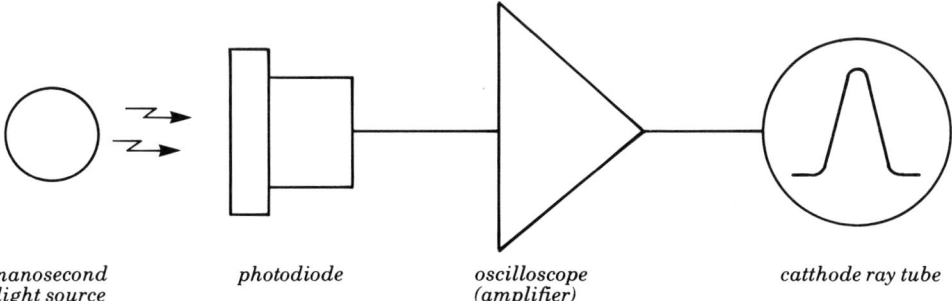

nanosecond   photodiode   oscilloscope   catthode ray tube
light source   (amplifier)

Figure 4.13 Risetime measurement by photodetector.

A plot of the output current of the detector versus the input radiation level there should show no change in the slope of the line from the lowest level of radiation to the highest level of radiation. Noise in the detector or system will determine the lowest level of incident radiation detectable. The upper limit of this input/output linearity characteristic is established by the maximum current capability that the detector can handle without becoming saturated (no change in output for a change in input). Linearity may be stated in terms of maximum percent deviation from a straight line over a range of input radiation levels.

Many factors may contribute to the deviations in linearity. Radiation levels in excess of the manufacturers stated maximum level may cause irreversible damage to the surface of the detector, saturation may be achieved, fatigue can occur, or hysteresis may appear. In the case of photomultipliers, temporary blindness is said to occur, and it may take a considerable time for the device to recover from this temporary overloading effect. A rating is usually given for the maximum allowable continuous radiation level. If the radiation is in the form of a very short pulse, it is possible to exceed the continuous level by a factor of ten or more without damage or noticeable changes in linearity.

"Quantum efficiency" may be defined as the ratio of countable events produced by the incident photons to the number of incident photons. Note that it does not include subsequent photons produced by amplification processes. As an example, in the case of a photomultiplier, quantum efficiency is a measure of the average number of photoelectrons emitted from the photocathode per incident photon at a particular wavelength. If, over a period of time, an average of 10,000 photoelectrons are emitted for 100,000 photons, then the quantum efficiency would be 10%. The quantum efficiency is equal to the responsivity times the photon energy of the incident radiation and is normally expressed as a percentage. The quantum efficiency is another way of measuring the effectiveness of the basic radiant energy in producing electrical current in a detector. It is interesting to note that the quantum efficiency of the dark-adapted human eye is a maximum of approximately 3% at a wavelength of 510 nm. The signal-to-noise performance of the eye is equivalent to that of an ideal detector which produces one recorded photoevent for each 33 incident photons within the wavelength band to which it is sensitive.

For most materials, the quantum efficiency is very low; on the best sensitized commercial photosurfaces, the maximum yield is as high as three photoelectrons for four light quanta. Exceptions are some types of bulk photodetectors called photoresistors and avalanche photodiodes which produce higher quantum efficiencies. The quantum efficiency of these devices may exceed unity; avalanche photodiodes are considered to be the fastest, most sensitive, broadband semiconductor photodetectors available.

The topic of "noise" is a broad subject and only the fundamental ideas are discussed as applied to photodetectors. Noise can be defined as any undesired

POWER AND ENERGY MEASUREMENTS

signal and can be divided into two broad categories: externally induced noise and internally generated noise. External noise, as the name implies, includes those disturbances that appear in the system as a result of an action outside the system. Two examples of external noise are pick-up from the 60-cycle power lines and static caused by electrical storms. Internal noise, on the other hand, includes all noise that is generated within the system itself. Every resistor produces a discernible noise voltage and every electronic device (the vacuum tube and the semiconductor) has internal sources of noise. The internal noise may be thought of as an ever-present limit to the smallest signal that can be handled by the system.

The first thing to note about noise is that it cannot be described in the same manner as the usual electric voltages or currents. It is common for us to think of a current or voltage in terms of its behavior with time. For example, we think of a sine wave as periodically varying with time, a direct current as being constant with time, etc. Now, if we look at the noise output of any electrical circuit as a function of time, it will be found that the result is completely erratic; that is, we cannot predict what the amplitude of the output will be at any specific time. Also, there would be no indication of regularity in frequency in the waveform. When completely unpredictable conditions such as this exist, the situation is described as random.

The individual must determine from preliminary noise measurements whether to make corrections to the photodetector instrument values.

Laser Power and Energy Detectors

Whenever a laser is applied for some specific purpose, the user generally will be concerned with the output energy or power of the laser for reasons such as the following:

Is the output strong enough to accomplish the intended purpose? For example, if a CO_2 laser is to be used for textile cutting, will a given laser cut the cloth when properly focused? Is the output so strong that it will harm the optical elements that it is used with, or constitute a safety hazard? For example, can the beam from a Q-switched Nd:YAG laser be manipuated by front-surface (silvered) mirrors, or will the silver be vaporized by the minute fraction of power absorbed by the coating? Is the intensity so high that if an accidental reflection causes the beam to enter the eye, will the eye be damaged? Is there an adequate power margin (or excess) to allow the system to operate in a slightly changed environment? For example, will a helium-neon communications link operate in a moderate haze? In rain? In fog?

All of the questions are concerned with how much power or energy is present at the output of a given laser. As mentioned previously, output of a continuous-wave laser is characterized by its *power*, measured in *watts*. The output of a pulsed laser is characterized by its *energy*, measured in *joules*. The joules emitted by a pulsed laser in a given pulse length determine its average peak power measured in watts, or joules/sec.

Two systems of units are commonly used to describe the strength and effect of optical radiation. The "radiometric" system is derived from basic physics and depends upon measurements of energy and power independent of the optics of the eye. The "photometric" system is concerned with the response of the human eye to different wavelengths. Photometric units are meaningful in situations where a laser output will be directly observed with the eye, such as a laser display.

Energy is defined as the ability to do work, for example, raising the temperature of a given amount of water a given number of degrees. A joule is a unit of energy that expresses how much work can be done by an energy source (such as a laser or charged capacitor). Power is the rate at which work is done. Power is expressed in watts, or joules per second. Many laser power and energy meters utilize these defined quantities as a basis for their measurement.

In addition to the total energy or power output from a laser, it is also common to be concerned with the "energy density" and "power density." The concept of density indicates how much area the energy or power covers. Energy density is normally expressed in joules/cm^2 and power density in watts/cm^2. Power density is referred to as irradiance.

Energy density or power density is calculated by dividing the total energy or power in the beam by the cross-sectional area of the beam. For a circular beam, the area would be πr^2 where r is the radius at the half-intensity points. It should be noted that the term "density" is usually concerned with the concept of *volume* (rather than *area*) in most other physical measurement terminology. Because these density terms are commonly used in electro-optics, the reader should be familiar with this concept.

Photometry is concerned with the relative response of a human eye to optical radiation as a function of wavelength. Therefore, photometric units are concerned with wavelength.

A curve of the response of the human eye to light as a function of wavelength is given in Figure 4.14. Note that the human eye has very little response except in the 425 to 675 nm region, and that the response is maximum near 555 nm.

Photometric units were intended to "normalize" the response of the eye to light of different wavelengths. In other words, the photometric system makes it possible to compare blue light of a given energy density to green light (or any other color) of the same energy density in terms of the ability of the eye to respond to the different wavelengths of light.

The "lumen" is the basic unit of power in photometry, much as the watt is the basic unit of power in radiometry. The significance of a lumen is that one lumen of light at 650 nm should appear as bright to the eye as one lumen at 555 nm.

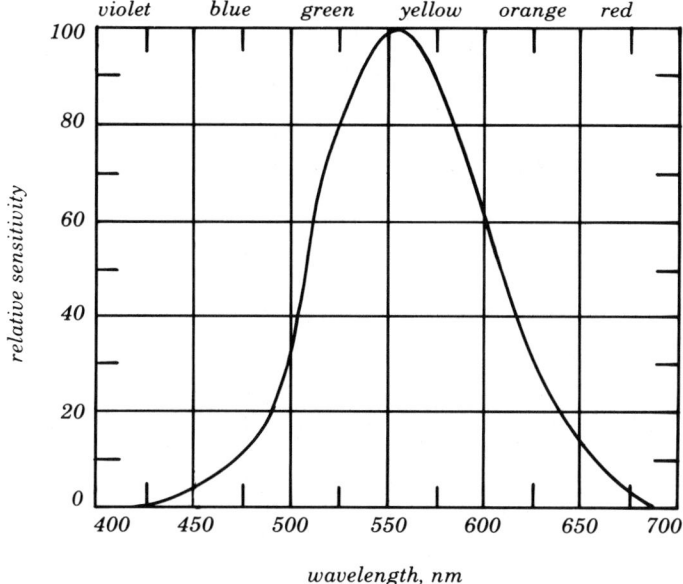

Figure 4.14 Relative sensitivity of the eye to visible radiation.

In contrast, 0.001 W of 555 nm radiation would appear approximately 10 times brighter (to the eye) than 0.001 W of 650 nm radiation. This can readily be seen from Figure 4.14 where the relative sensitivity of the eye at 650 nm is about 1/10 that at 555 nm.

The importance of the lumen, therefore, lies in comparing the eye's response to light of different wavelengths. Consequently, the application of photometric units to laser technology is limited to systems that will be directly viewed by the human eye.

The common methods of measuring the energy or power output of a laser are based upon observing the effects of the intercepted beam on a device or material that is changed by the incident radiation. (A power meter is shown in Figure 4.15.) The type of change may be

1. Heating, as in a calorimeter
2. Photoelectric or quantum effects, as in photodetectors
3. Photochemical effects, as occur when photographic film is exposed to optical radiation

These three techniques will be briefly described, and their application to the laser sources will be discussed in greater detail.

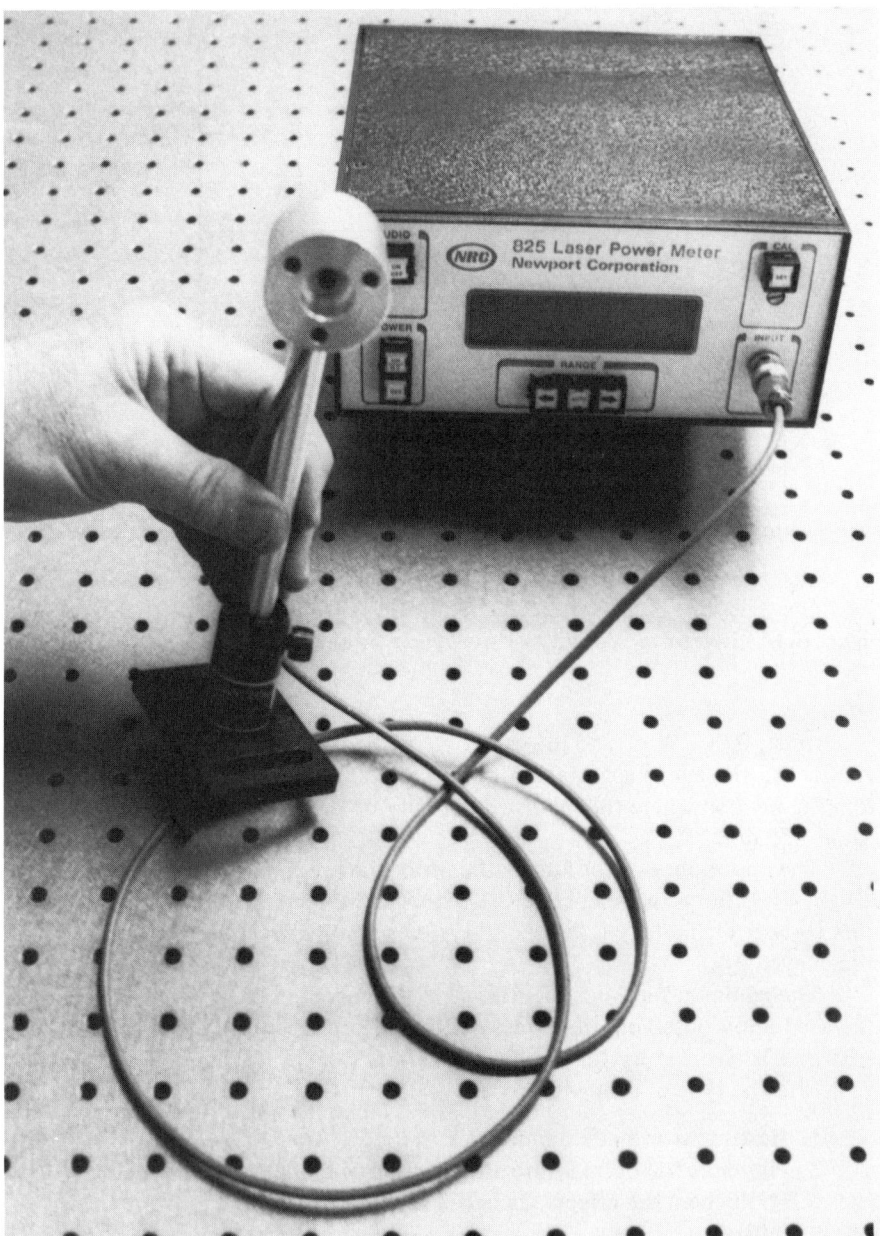

Figure 4.15 Photograph of a laser power meter assembly. Laser beam is directed into hand-held wand and the meter displays digital readout. (Courtesy Newport Corp.)

POWER AND ENERGY MEASUREMENTS

A calorimeter normally consists of (a) a target which receives, absorbs, and contains the incident radiation, and (b) a thermal sensor to indicate how much temperature rise has occurred. Various schemes have been developed to optimize the geometry of the absorbing structure by using multiple reflections and special coatings. Care is taken to prevent loss of heat to the environment by drafts or contact with thermally conductive materials.

The calorimeter output is a function of

1. The absorbed energy
2. The mass of the absorbing material
3. The specific heat of the absorbing material
4. The sensitivity of the thermal sensor
5. Various losses which limit the heat rise of the target (such losses as partial reflection of energy, vaporization of material, convection losses, re-radiation losses)

The more common methods of temperature sensing consist of thermocouples, thermistors, and resistance-wire bolometers. Thermocouples, deliver a voltage that is a function of temperature for a given metal-to-metal junction. Thermocouples can be calibrated very accurately but are relatively insensitive, and care must be taken to measure accurately the thermocouple output voltage. Frequently, laser calorimeters use series connections of thermocouples to increase their sensitivity; these series connection of thermocouples are called thermopiles.

Thermistors are semiconductor devices which are characterized by a temperature-sensitive resistance. Thermistors are more sensitive than thermocouples but less accurate and prone to change with age.

Calorimeters used to measure energy are constructed so that the heat losses to the environment (during the time that the measurement is being made) are small compared to the energy that was absorbed. Calorimeters used to measure power are constructed so that the absorbed heat is transferred to the environment at a known rate. Power meters for higher power lasers are constructed with cooling fins or even water cooling to achieve their transfer to the environment. If the heat-transfer characteristics of the absorber assembly are accurately known, the power meter can be calibrated from a knowledge of the rate of energy arrival and transfer when the system reaches thermal equilibrium (a situation in which the temperature is neither rising nor falling with a fixed input).

Calorimeters sometimes contain a resistance wire in direct contact with the absorbing material such that a known amount of energy can be introduced from some well controlled electrical source. The indicated meter reading for this known input is then used to calibrate the meter response.

Calorimeters are generally insensitive to wavelength of the incident radiation; because of their thermal, rather than photoelectric, nature, they are useful from the ultraviolet to the infrared spectra.

Any device that has a photoelectric sensitivity to electromagnetic radiation of a given wavelength can be used as a power or energy measuring device for that wavelength. In contrast to calorimeters, which respond over a broad optical spectrum, most "photodetectors" are wavelength sensitive, and measuring devices which utilize such detectors must be corrected for the wavelength of the source being measured. A calibration curve describing the variation of sensitivity as a function of wavelength usually accompanies commercial instruments which utilize photodetectors. For measurements on monochromatic sources, the use of the calibration curve is a simple matter of finding the proper wavelength on the X axis of the curve, reading the corresponding correction factor on the Y axis, and multiplying the correction factor by the meter reading.

The commonly used photoelectric power meters utilize silicon photodiodes, photomultipliers, and cadmium sulfide photocells. The limitations of the power meters are essentially the same as for the photodetector utilized in the meter. Instruments that measure radiation in the ultraviolet are not ideally suited for infrared measurements, and high-sensitivity photomultiplier-based instruments are not ideally suited to measuring the output of a 100-W CW YAG laser. For these reasons, some power- and energy-measuring instruments are constructed of modular systems which can be configured for a particular measurement. For example, an instrument may have several replaceable detector head assemblies which cover different spectral regions with both high- and low-sensitivity options.

A number of techniques have been developed which indirectly indicate energy or power by physical interaction of the electromagnetic radiation with certain materials. For example, nonlinear crystals, photographic film, gas expansion chambers, liquid expansion chambers, and phosphorescent chemical films have all been used in the past for measurement of the intensity of optical radiation in specialized cases. These techniques tend to be difficult to calibrate and subject to errors in reproducibility. They are rarely encountered in commercial instruments but may be useful in special applications where calorimetric or photoelectric techniques are not appropriate.

A block diagram of a typical modular-construction power/energy meter is shown in Figure 4.16. It is apparent from the diagram that a number of accessories are generally used with the photodetector to assemble a power-measuring instrument with the versatility to handle a variety of optical sources. We are concerned with measuring sources with the following widely varying parameters.

1. Wavelengths that may vary from 200 nm to over 10,000 nm
2. Power levels that may vary from 10^{-12} W to 10^3 W CW, or above 10^{-9} W peak
3. Energy levels from 10^{-6} J to 10^3 J
4. Pulse widths from 10^{-9} sec to 10^{-1} sec, and CW sources

The functions of the components in Figure 4.16 and their effects on these four parameters will be individually discussed.

POWER AND ENERGY MEASUREMENTS

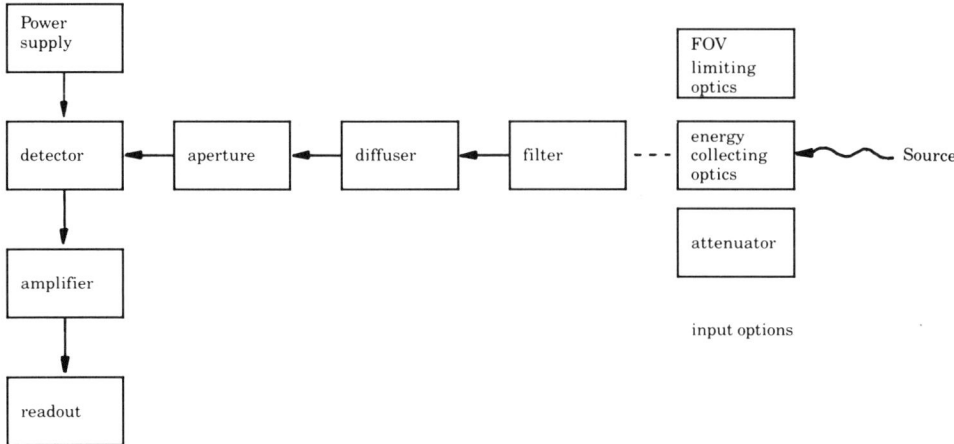

Figure 4.16 Block diagram of a typical power/energy meter arrangement.

Input Optics

Accessories at the input to the power meter function to (a) gather more energy, particularly in situations where the divergence of the incident radiation is large; (b) reduce the intensity of the incident radiation by a precisely known amount in situations where the intensity is so high that damage to the instrument would occur, and (c) limit the field of view of the instrument, such as in situations where the background radiation is almost as strong as the source radiation to be measured.

Corrections to the calibration of the instrument must be made when any of these accessories are used.

If an energy-gathering lens is used, the relationship between actual power received by the detector (or energy) and indicated power received by the detector is given by:

$$\text{actual power} = \text{indicated power} \times \frac{\text{detector input aperture diameter}^2}{\text{lens diameter}^2}$$

provided that all the radiation gathered by the lens enters the detector aperture. When very precise measurements are necessary, a further correction must be made for the transmission and reflection losses of the lens at the wavelength of interest.

If an attenuator is used, the relationship between indicated power and actual power will be:

$$\text{actual power} = \text{indicated power} \times \frac{1}{\text{attenuator transmission}}$$

Care must be used when attempting to use attenuators to reduce the output of high peak power devices, especially Q-switched lasers. It is possible for the coating on the attenuator to be vaporized and the unattenuated laser output to damage the power meter. In beam-splitter-type attenuators, the user should be aware that the deflected portion of the laser energy must be treated with the same respect as the laser output; it can do almost as much damage. Beam splitters are also polarization sensitive, and it is always desireable to calibrate the attenuation afforded by reflection beam splitters for the given source under measurement. This can most easily be done through the use of two detectors at a laser optical output level. Absorption-type attenuators (such as Wratten filters) can be used at lower power levels. These filters are wavelength sensitive, and corrections are necessary if they are used outside the visible region.

Some commercial power meters have a telescope option available which limits the detector field of view to a very narrow angle, thus rejecting ambient light that would interfere with the measurement of a low-level source. In most cases, the telescope objective lens is larger than the input aperture of the power meter, so that a correction of indicated meter reading is necessary. This information is usually supplied by the manufacturer of the power meter.

Spectral Filter

A spectral filter may be included to limit background radiation or to flatten the detector's spectral response, as discussed previously. If a spectral filter is used, the spectral characteristics of the source and filter must be known and corrections made as described in the section on detector response.

Diffuser

It is advantageous to illuminate the entire sensitive area of a photodetector for two reasons.

1. Damage or nonlinearity caused by focusing too much energy into a small spot is avoided
2. Photosensitive surfaces usually have variations in responsivity that are minimized by illuminating a larger area

The diffuser has the property of spreading the incident energy but absorbing or reflecting very little of it. Diffusers are usually made of ground or opal glass, a milky colored glass that has a high damage threshold. The diffuser also minimizes alignment sensitivity of the power meter.

Aperture

An aperture of variable diameter placed in front of the detector allows the measurement of a wide range of power or energy levels with the same detector. If the diameter of the aperture is accurately known, the attenuation resulting from

POWER AND ENERGY MEASUREMENTS

the aperture can be easily calculated and the actual power calculated from the indicated power.

Detector

The detector may be either a photoelectric or calorimetric device; the photoelectric type is more common. Several types of detectors may be available to accommodate different wavelength regions and different source intensities. Photomultiplier tubes and photodiodes are the most commonly used detector elements.

As an example of how a photodetector can be used to measure both power and energy, let us consider a back-biased silicon photodiode detector head. A photodiode has a measured responsivity of 0.400 mA/mW of monochromatic radiation at 600 nm and has a risetime of 5 nsec. To measure continuous W and CW power, we need only to measure accurately the current flowing through the diode when the source is incident upon the detector surface. Then

$$\text{Power (mW)} = \frac{\text{photocurrent (mA)}}{0.400 \text{ (mA/mW)}}$$

This same relationship holds for pulsed-mode peak-power measurements so long as the rise time of the detector is small compared to the risetime of the pulse.

The second means of estimating the energy in a pulse is to allow the resulting current pulse (from the photodiode) to charge a capacitor of known value. The amount of charge which builds up on the capacitor is related to the energy in the pulse by:

$$E = \frac{CV}{0.400 \text{ mA/mW}}$$

where

E is energy in joules
C is capacitance in farads
V is voltage across the capacitor in volts

In practice, this technique is more complex. The voltage across the capacitor must be read when it has reached its maximum value but before the capacitor has begun to discharge. A special circuit, which places a very high impedance (several hundred megohms) in series with the capacitor, keeps the charge on the capacitor from discharging while the voltage is being measured.

Power Supply

The power supply furnishes the voltage for the photodiode. If a photomultiplier is used, a highly regulated DC supply is necessary because of the extreme sensitivity of current gain to diode voltage.

Amplifier/Detector Load

This subsystem converts photocurrent to voltage. In some cases, a precision resistor may be an adequate conversion device. If linearity over a broad range of input conditions is necessary, it will be advantageous to use an operational amplifier as the photodetector load. Because the detector is then operating into a very low impedance, linearity is enhanced.

The value of the load resistor is chosen on the basis of output voltage and risetime considerations. Several selectable load resistors may be present to facilitate the selection of the optimum load.

Readout Device

The final subsystem in the measurement scheme is a meter movement or recorder for monitoring the amplified photocurrent. In some systems, this may take the form of a digital readout. Various scale-changing options may be included, so that differing intensities can be measured by simply turning a rotary switch. An adjustment to null out ambient radiation is also usually included.

PHOTOGRAPHIC INSTRUMENTATION

The term "photographic instrumentation" is the use of a photosensitive medium for the detection, recording, and/or measurement of scientific or engineering phenomena.

Numerous types of photographic systems exist which fall within the above definition. Only a brief discussion of the accessories and techniques of oscilloscope photography in the recording of repetitive and nonrepetitive events will be presented, in addition to the basic operation of the single-lens reflex camera and brief discussion of the characteristics and applications of image converters, image intensifiers, as well as some types of high-speed camera systems useful for the observation of laser-induced phenomena.

In order to compare various photographic instrumentation systems and suitable various photosensitive recording material, it will be necessary to define several terms useful in comparing and selecting alternate recording devices and materials. Included will be a description of how various factors affect the optimum recording of oscilloscope traces. The concept of "writing speed" and the influence of recording material parameters on this value will also be presented. Camera systems have an effect on writing speed but will not be discussed here. However, a short description of high-speed motion analyzers, and novel electro-optic photographic systems such as image converters, will follow.

"Writing rate" is the ability of a particular camera system mounted on a particular oscilloscope to photograph fast-moving traces. The writing rate figure expresses the maximum spot rate (usually in centimeters per microsecond) which can be photographed satisfactorily.

The faster an oscilloscope spot moves, the dimmer the trace will become. This is because the moving electron beam strikes each point on the phosphor coating for a shorter period of time. A camera-recording system and high-speed oscilloscope are required for photographing low repetition rate displays at fast oscilloscope sweep rates. Absolute writing rate of an oscilloscope or camera system is difficult to measure because of the many variables involved, such as the speed of the lens (f-number), the type phosphor used on the cathode ray tube, sharpness of electron beam focus, the type of film used, development time, and beam intensity. Some of the factors that affect the writing rate are film variables.

The American Standards Association (ASA) designated speed of a film reveals little about its effectiveness in recording single oscilloscope traces. This is because the ASA speed rating is assigned to the film when exposed for 1/50 second to light of normal daylight spectral characteristics, compared to the very short exposures of fast traces, which can be several orders of magnitude smaller and have much different spectral distributions. In photographing fast traces with an oscilloscope, either Polaroid 410 film with an ASA rating of 10,000 or Polaroid type 107 or 47 with an ASA rating of 3000 is used. Type 410 film appears to have a higher speed rating by a factor of greater than three. The resolution for both of these films is approximately 22-28 lines/mm so the difference in resolving power is negligible.

Writing rate is greatly affected by the type of phosphor used in the cathode ray tube. A good combination for recording high-speed pulses is matching a type P-11 phosphor to type 410 film. This gives a very high writing rate.

The writing speed of Polaroid films used in oscilloscope photography can be increased (through hypersensitization) by a factor of 3 for type 47 (3000 speed) roll film and type 107 (3000 speed) pack film, while the writing speed of type 410 (10,000 speed) roll film can be increased by a factor of 2.

"Prefogging" can be described as a process of exposing the film to a burst or short exposure of low-level red light essentially to raise its sensitivity level before it is exposed to the trace to be photographed on the oscilloscope screen. A device called a writing speed enhancer, which is attached to the oscilloscope camera, can be used to control accurately the amount of prefogging. The film can be fogged before the sweep occurs, after the sweep occurs, or at the same time the sweep occurs. These techniques are called prefogging, postfogging, and simultaneous fogging, respectively. It has been found that simultaneous fogging provides the greatest gain in writing speed. Decreasing the development time will also enhance the writing rate, but with a loss of contrast.

"Transillumination" is a method of effectively increasing the writing rate. It permits better viewing of information which is recorded primarily on Polaroid prints. In transillumination, the print is observed with a source of bright diffused light, such as a tungsten light bulb, directly behind the print. The light passing through the print brings out detail which would not otherwise be evident. The enhancement in writing rate so produced is approximately 20% over that of direct viewing. Polaroid type 47 or 410 film can be used in this manner but not 107, which is opaque.

High-speed cameras for recording ultrafast events are based on a continuous motion of the film or by the image being scanned without mechanical shuttering. If a single photographic record of an event is taken by means of a short exposure of light, such as a stroboscope, flash tube, or by a fast mechanical or electro-optical shutter, we see only one incremental stage of the total motion under observation. Essentially what we observe is a single short period of time, and we have no record of what preceded or followed the changing event. "Multiple-framing cameras" and "streak cameras" permit chronological measurements of events that occur at speeds beyond the limit of human visual perception. We know from experience that if an event takes place too rapidly, only the effects occurring before and after are visually recorded. As a general rule, the human eye requires about 10 msec to perceive a direct stimulus, and it retains the sensation for approximately 50 msec after the stimulus is removed. Two rapidly occurring events spaced closer than approximately 1/40 second may be perceived as one. The term "critical flicker frequency" is the frequency above which the eye is incapable of detecting individual events and varies with the amount of illumination.

The multiple-framing camera record consists of a sequential series of two-dimensional images of a three-dimensional object, so the X, Y, and Z coordinates of the object are reduced to X and Y coordinates on the two-dimensional film as a function of distinct time intervals. Time resolution is limited to the time between frames.

The streak camera records spatial information in the Y coordinate as a function of continuously written time, but no information in the X coordinate. Streak records are commonly made for recording wavelength versus time (streak spectrograph) and also displaying intensity of the event as a function of time.

Image dissection techniques will yield a continuous display of time but discontinuous spatial recording in both the X and Y coordinates. In each of these methods, the film density of the image is a function of the intensity of the event being recorded.

A very popular type of high-speed framing camera is the Dynafax, which is a rotating drum-type of camera capable of producing 224 16-mm size frames at framing rates between 200 to 35,000 per second at 0.7-, 1.8-, and 3.7-μsec shuttering speeds on a 33-inch length 35-mm film. Pictures are spaced in two rows along each side adjacent to the film perforations. Operation is based on a rotating drum as a film transport and an octagonal rotating mirror for shuttering. The camera is gradually brought up to speed and stabilized before exposures are made.

Speed is controlled by a variac and monitored by a built-in tachometer calibrated in frames per second.

In the operation of a streak camera, the event can be focused on a slit of the camera such that the slit is perpendicular to the direction of motion of the event but, of course, parallel to the spatial dimension of the event to be recorded. In this way, a record is obtained in which the measurement along the strip of film is time, and the measurement across the film strip is spatial information. Such a record is called a "streak" or "smear" photograph. If a grating is employed along with a slit, we have what is called a time-resolved spectrograph record of wavelength versus time.

A method for avoiding the very high-speed motions required in framing and rotating mirror cameras is found in a system called "image dissection." In this method, the image plane is subdivided into a large number of elements by means of a grid, a lenticular plate, or similar apparatus, and the dissector is moved a relatively short distance during the exposure. The result is a series of exposures of small elements of the scene. Application of fiberoptics permits very fine dissection and a rearrangement of the dissected elements into a different array, such as a straight line from a two-dimensional image. It is interesting to note that a combination of an image-dissector and image converter is presently capable of taking a series of photographs in excess of 10^9 pictures per second. Exposure times have now been reported to be in the subpicosecond range using laser light or optically driven Kerr cells.

Applications of these types of cameras are too numerous to list, but one can simply state that high-speed cinematography is dedicated to one purpose: the recording of fast recurring or transient events which cannot be studied in detail with conventional photographic methods.

An image converter tube is an essential element in an "infrared telescope," and it serves to convert the invisible infrared image into a visible image. The tube contains a semitransparent photocathode, which is sensitive to infrared radiation, and an electron lens which images the electrons from the photocathode onto a fluorescent screen. When an infrared image is focused on the photocathode, a visible reproduction of this image is formed on the fluorescent screen.

Basically, the infrared telescope consists of the image tube, an objective for forming the infrared image on the photocathode, and an ocular for viewing the reproduced image. Associated with the telescope is a battery-operated, vibrator power supply which furnishes the 4000 to 5000 DC voltage and several intermediate voltages required by the image-converter tube.

Several types of telescopes have been developed and produced for a number of different applications. These include a signalling telescope employing a large aperture reflective optical system as objective, the "sniperscope" which is a carbine-mounted telescope and infrared source permitting aiming and shooting in complete darkness, and the "snooperscope," composed of the same infrared units mounted on a handle for short-range work. The four basic components of the complete sniperscope are the telescope, vibrator-type power supply, a tungsten light source with infrared filter attached, and the telescope handle. The principle of operation of this type of sniperscope is presented. The human eye sensitivity extends between about 0.4 and 0.7 microns. An ordinary tungsten light is the optical source. The cesium oxide photocathode is sensitive in the visible and infrared regions up to about 1.2 μm, and it is, therefore, sensitive through an appreciable region beyond that of the human eye. A filter is placed over the tungsten light source, eliminating the radiation emanating from the source within the visible region. The band in which the radiation from the filtered tungsten light source overlaps the sensitivity of the cesium oxide is the useful radiation employed to illuminate the target or object under surveillance. The image observed through the eyepiece is green, corresponding to the spectral output of the fluorescent screen and has the same tonal variations as the object that is being viewed. Nominal resolution of such infrared viewers is in the order of 10 to 14 line pairs per millimeter.

Operation of the sniperscope (or snooperscope) is quite simple. There are two controls found on the power supply unit. One is the on-off switch and the other is labeled "electrostatic focus." The objective lens is adjusted along with the eyepiece to obtain the sharpest image. Some adjustment of the electrostatic focus may be required. The tungsten lamp is operated from a switch located on the trigger handle. All adjustment should be made while the lamp is energized.

An "image-intensifier night-vision device" uses several electro-optic devices similar to those found in the image converter, but with the addition of multiple units of fiberoptic arrays in order to obtain light amplification. Unlike the older sniperscopes which required a separate infrared light source, these devices are entirely passive, operating from whatever natural sources of light are available. Sometimes referred to as a starlight scope, its operation is described. The very low irradiance image is focused onto the first photocathode by the objective lens which results in photoelectrons being emitted in proportion to the incident radiation. These electrons are accelerated by a 15-kV electrostatic field to the first phosphor screen, where the energy of the electrons is transferred back into visible light. This process is repeated in each of three stages until a significantly intensified image is formed at the exit of the eyepiece. The multiple arrays of fiberoptics help to transmit the image from each phosphor screen to its adjacent photocathode without excessive distortion or attenuation. Total light gain is on the order of 5×10^4.

Applications of image-converter tubes other than military are numerous. Among these are the following: police surveillance, medical observations, nocturnal animal behavior, opthalmoscopy, spectroscopy, and alarm systems. There are numerous scientific applications as well.

The basic elements of a "high-speed image-converter camera" are the objective lens which focuses the light radiated from the event under observation onto a semitransparent photocathode of the image-converter tube and the photocathode which transforms the photon image into an electron image. Shuttering, frame displacement (deflection), and amplification are accomplished electronically on these liberated electrons. The electron image is focused to cross between the deflection plates and is imaged on the photocathode where it is converted into a higher intensity photon image and relayed to the film through a double-coated lens system. Distortion and edge resolution problems have been minimized by treating the objective lens and image converter as a system. The photocathode of most image converters is curved to provide virtually uniform resolution across the picture. The objective lens introduces a field correction making the spatial (x-y) distortion of the system negligible.

The gating grid in the image converter tube serves as the ultrahigh-speed electronic shutter; electrons flow only when it is pulsed. Shutter-open to shutter-closed light contrast is better than 10^6. Response time from trigger signal to the grid to the start of recording is typically 20 nsec, with exposure times as short as 5 nsec.

Some types of high-speed image converter cameras are capable of operating in four different modes: (a) as a framing camera in which individual records of an event taken at different times are recorded with the capability of recording multiple frames in which the exposure time of each frame and the delay between frames may be varied; (b) as a streak-operated camera providing a continuous written record of time and spatial information along one axis; (c) as a stroboscopic camera for a single frame of a variable time exposure; and (d) as a time-resolved spectrograph to display time versus wavelength. In each of these modes of operation, the brightness of the image recorded on the film is related to the intensity of the event. Unfortunately, these types of cameras do not have a large dynamic range and will saturate if overloaded with intense light.

While image converter cameras are normally triggered by the event to be photographed, the camera trigger circuitry may cause a delay before the camera is actually gated on. To photograph the beginning of an event, therefore, it is necessary either to obtain an earlier trigger or to introduce an optical delay into the event itself. If the event is a laser spike, an optical delay can be easily obtained by taking the trigger pulse from a beam splitter in the laser or near the laser itself. The image can be delayed by increasing the path length from the laser output through a series of mirrors before impinging onto the photocathode of the image converter tube. Accessories and detailed specifications are too numerous to list here, but, in general, it may be said that the image-converter camera is an ultrahigh-speed device with the capability of providing frame exposure durations from 5 nsec to 10 μsec; streak writing rates from 5 mm/msec to 2500 mm/msec;

and stroboscopic rates up to 30 kHz with exposure durations as short as 5 nsec for observing luminous or nonluminous events.

Applications include observation of explosive detonations, plasma studies, axial mode beating in lasers, study of corona discharges, hyperballistics, shock tube phenomena, and spectroscopy.

BIBLIOGRAPHY

Jenkins, F.A. and H.E. White, (1957, 1976). *Fundamentals of Optics*, McGraw-Hill, New York.

Jones, J., J. Ready, et al. (1972). "Course 7, Laser/Electro Optic Devices, Laser Electro Optics Series," Technical Education Research Centers, Inc., Cambridge, MA.

Pedrotti, L.S. and H. Weichel, (1972). "Course 3, Laser/Electro Optic Measurements, Laser Electro Optics Technology Series," Technical Education Research Centers, Inc., Cambridge, MA.

5
Safety in the Laser Environment

Many users of lasers do not desire an extensive understanding of the physics and mathematics of lasers, nor an in-depth description of biological damage mechanisms; they may have a limited technical background and are adapting the laser to specialized applications. The inability to grasp technical details, however, may lead to carelessness, needlessly increasing the danger to operating personnel.

Laser hazards can be minimized by a few simple engineering and administrative controls. Proper indoctrination and training of laser personnel in limited aspects of laser hazard control is, therefore, very important.

If orientation in laser safety includes the fundamental requirements and philosophy of American National Standard Z136.1-1986, *Standard for the Safe Use of Lasers* (1), and impresses upon those working in the laser environment what the minimum biological damage threshold values are, then the users will understand the hazards, will follow reasonable control procedures, and will be comfortable working with this valuable emerging technology.

The details described here include several practical applications of laser hazard controls developed at Los Alamos, where no permanent biological damage had been reported by laser light during more than ten years experience in applications of a large variety of lasers and laser systems.

LASER BEAM CHARACTERISTICS

To control the hazards of any laser, one must know the wavelength, the maximum output power (if *continuous wave*), or energy (if *pulsed*), and the diameter of the emanating beam. Pulse width and pulse repetition rates also are important. Next to the wavelength, however, the most critical information about each laser is its *power density*, if continuous wave, and its *energy density*, if pulsed.

Wavelength is vital information because it establishes whether the beam is focusable by the eye. The wavelength vs. transmission curve for the eye is shown in Figure 5.1. If the wavelength is between 0.4 and 1.4 μm, the human ocular system focuses the incident beam by as much as 100,000 times on the retina. This wavelength region is called the ocular focus region, or retinal hazard range. The visible portion of the ocular focus region in which the eye detects color ranges only from approximately 0.4 to 0.7 μm. Wavelengths in the range 0.7 to 1.4 μm are not detected by the retina; they are invisible to the ocular system, although they are focusable by the eye. The popular Nd:YAG laser emits a wavelength (1.06 m) in this invisible portion of the focusable range, yet this wavelength has very low retinal damage threshold values (see Fig. 5.2).

Lasers with wavelengths within the ocular focus region require more control, because much less energy accessible to the eye is required to cause damage.

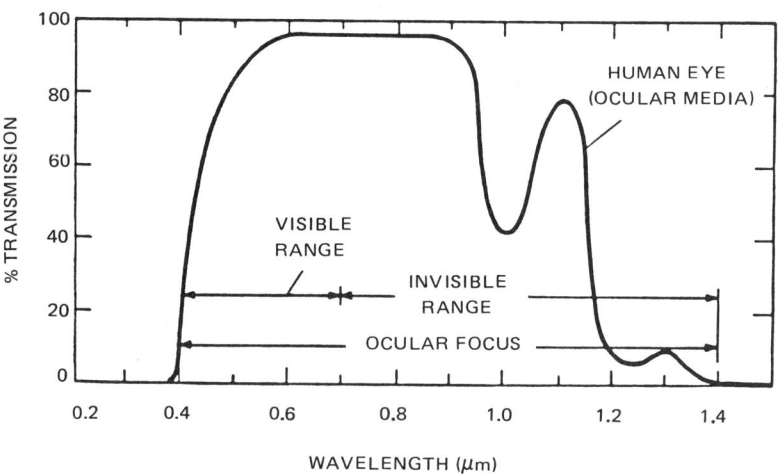

Figure 5.1 Transmission curve for the human eye showing visible and invisible ranges within the ocular focus region of the spectrum.

LASER BEAM CHARACTERISTICS

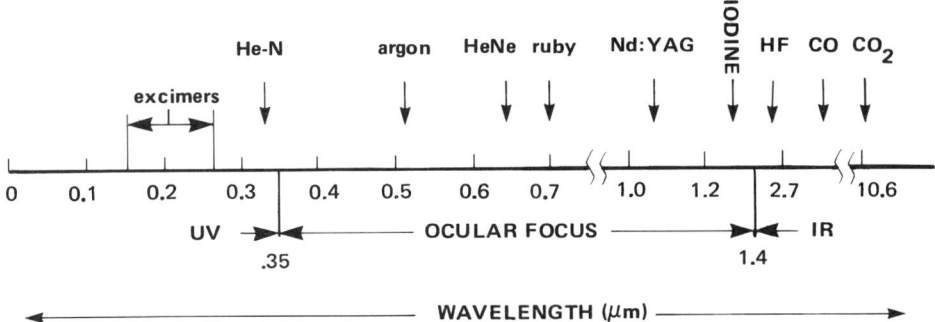

Figure 5.2 Spectrum locations of some of the more popular laser beam wavelengths.

Continuous wave (CW) lasers cause biological damage mainly by thermal mechanisms, resulting in burns or blistering from the steady beam. If the wavelength of the beam is in the ocular focus region, eye damage occurs in the retinal tissues, because very little energy is absorbed by the cornea, lens, and aqueous tissue. Wavelengths outside the focusable region are absorbed by the outer components of the eye, especially the cornea (see Fig. 5.3).

The skin absorbs all laser wavelengths, but much more energy is required for skin damage than for eye damage, and more energy from CW lasers is required for damage than from pulsed lasers. If a laser emits radiation continuously for a minimum of 0.25 seconds, it is considered a CW laser.

Pulsed lasers release beams in short bursts or pulses. Biological damage from fast-pulsed lasers in caused mainly by shock-wave or blast mechanisms not unlike those caused by a bullet. Much less energy is required from pulsed lasers than from CW lasers to cause eye or skin damage. The shorter the pulse, the greater the hazard. Fast repetition of pulses adds to the hazards.

Beam diameters determine the "densities" of energies emerging from lasers. It is the energy density that is important in hazard control. Each commercial laser manufacturer is required to specify the diameter of the instrument's beam when applying for certification; this property is listed in the *Laser Classification Guide* (2), published by NIOSH; it contains the characteristics of all certified commercial lasers manufactured prior to July 1976. Any change in beam diameter or output beam characteristics must be recorded by the user and considered in any safety analysis of the laser's environment.

Power density is calculated from the size of the beam and the output power of a CW laser. It is expressed in watts per square centimeter (W/cm^2).

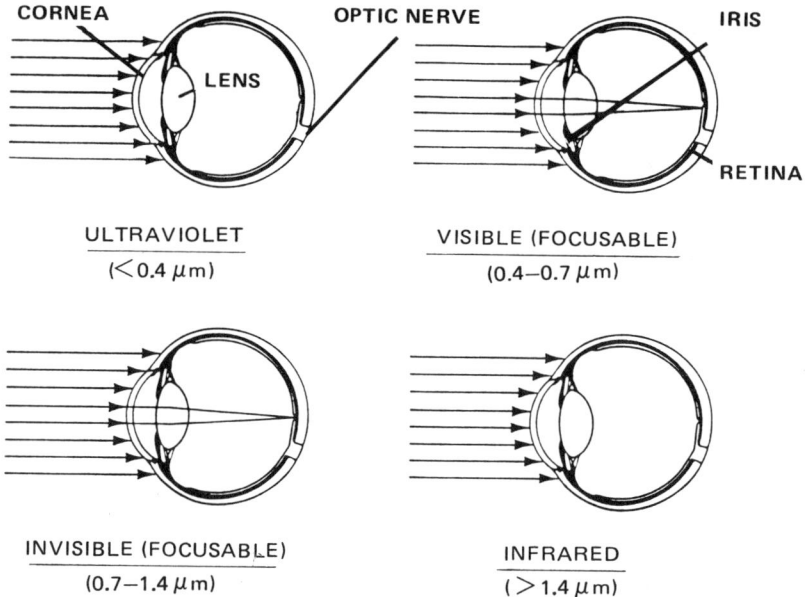

Figure 5.3 Explanation of which wavelengths are absorbed by eye components.

Technically, this property of a beam is called irradiance by physicists and engineers, but such words as radiance, radiant energy, or irradiance (fundamental terms to physicists) are confusing to many nontechnical users; the simple terms *power density* for CW lasers and *energy density* for pulsed lasers are acceptable, understandable terms for nontechnical users of lasers.

Power, in watts, is expressed as one joule per second (W-J/s); the shorter the duration of time (s), the greater, therefore, the power. For example, one joule pulsed at 10^{-3} s is 1000 W. One joule pulsed in 10^{-6} s is 1,000,000 W, and one joule pulsed in 10^{-9} s is one billion W, etc.

Energy density is calculated from the size of the beam and the output energy of a pulsed laser. For pulsed lasers, the pulse width, or temperal length of the pulse, measured in seconds, also must be known to determine damage threshold values. The beam intensity is technically termed radiant energy, but the term energy density is acceptable; it is expressed in joules per square centimeter (J/cm^2). This value divided by the pulse width in seconds is the beam's *power density*.

Pulse repetition rate, pulses per second (pps) is self-explanatory but is an important property in the evaluation of biological damage threshold values. Recently completed biological studies with the repeated fast pulses indicate that

BIOLOGICAL DAMAGE THRESHOLDS

threshold values generally are lower, depending on the rate, than for single pulses. Repetition rates slower than one pulse per second are considered single-pulse lasers.

BIOLOGICAL DAMAGE THRESHOLDS

Biological damage threshold values for the several categories of lasers are listed in Figure 5.4.

These are values above which permanent damage from direct viewing or from mirror-like (specular) reflection is likely to occur; they are recommended to be used as the basis for determining the engineering controls and eye protec-

Available intensities	UV .2–.3 µm	Ocular focus .5–.7 µm	1.1 µm	IR 2.7 µm	10.6 µm
CW (W/cm^2)^b	3×10^{-3}	4	28	$(3)^a$	3
pulsed (J/cm^2):					
100 nsec	(300×10^{-3})	110×10^{-3}	2.2	300×10^{-3}	(300×10^{-3})
30 nsec	(200×10^{-3})	(100×10^{-3})	(2.0)	(200×10^{-3})	(200×10^{-3})
20 nsec	(200×10^{-3})	(100×10^{-3})	(2.0)	(200×10^{-3})	(200×10^{-3})
1 nsec	(200×10^{-3})	(100×10^{-3})	(2.0)	(200×10^{-3})	230×10^{-3}
30 psec	(200×10^{-3})	(100×10^{-3})	(2.0)	(200×10^{-3})	(200×10^{-3})

[a] All parenthetical values are estimates by the author.
[b] Assume one second maximum exposure.

Skin damage threshold values for laser radiation (Caucasian).

Available intensities	UV .2–.3 µm	IR 2.7 µm	10.6 µm
CW (W/cm^2)^b	10×10^{-3}	$(100 \times 10^{-3})^a$	100×10^{-3}
pulsed (J/cm^2):			
100 nsec	(4×10^{-3})	4×10^{-3}	(5×10^{-3})
30 nsec	(4×10^{-3})	(4×10^{-3})	(5×10^{-3})
20 nsec	(4×10^{-3})	(4×10^{-3})	(5×10^{-3})
1 nsec	(4×10^{-3})	4×10^{-3}	5×10^{-3}
30 psec	$(<4 \times 10^{-3})$	$(<4 \times 10^{-3})$	$(<5 \times 10^{-3})$

[a] All parenthetical values are estimates by the author.
[b] Assume one second maximum exposure.

Approximate damage threshold values for corneal tissue for pulsed and CW lasers (minimum reported).

Available intensities	Ocular focus .5–.7 µm	1.1 µm
CW (W/cm^2)^b	10×10^{-3}	$<5 \times 10^{-3}$
pulsed (J/cm^2):		
100 nsec	$(100 \times 10^{-6})^a$	(100×10^{-6})
30 nsec	70×10^{-6}	(100×10^{-6})
20 nsec	130×10^{-6}	105×10^{-6}
1 nsec	(120×10^{-6})	(100×10^{-6})
30 psec	18×10^{-6}	9×10^{-6}

[a] All parenthetical values are estimates by the author.
[b] Assume one second maximum exposure.

Approximate damage threshold values for retinal tissue, pulsed and cw lasers (minimum reported).

Figure 5.4 Representation of lowest reported biological damage for various wavelengths.

tion in each laser's environment, using maximum permissible exposures (MPE) values in ANSI Z136.1 as a guide.

The ANSI Z136.1 *Standard for Safe Use of Lasers* does not list damage threshold information. Instead, it lists the MPE, which are related to biological damage and are "... (sometimes as much as a factor of 10) below known hazardous levels," as stated in Section 8 of that document. No explanation is given as to how these numbers are determined, nor is it convenient to interpret MPE data listed in the Standard's Tables 5 and 6. Also, some recent experimental data are not included (3).

Knowing the biological damage threshold values and the MPE data for the various lasers in terms of power density or energy density, laser personnel should be protected from any beams above these values. This approach not only will help prevent accidents but also will conform to the intent of any regulation or standard.

ENGINEERING CONTROLS

The ANSI Z136.1 standard requires the use of engineering controls for Class 4 lasers "... capable of producing hazardous diffuse reflections." However, damage can result from specular, or mirror-like, reflections from laser beams of lower energy density than listed in the standard as not requiring controls. Users can be lulled into a false sense of security by strict adherence to the standard's classifications and expect to be fully protected. For example, the standard required no control measures for Nd:YAG (1.06 μm) pulsed (10 ns) lasers emitting up to "31 $\times$ 10^{-3} J/cm^2," yet one documented exposure (4) to a suspected specular reflection from a 6 $\times$ 10^{-3} J/cm^2 (a factor of five less) resulted in permanent retinal damage. The limits for classification of CW lasers in the standard are expressed in watts and not in power density (W/cm^2), so that guidelines for CW lasers are not possible unless fairly complicated calculations are made, requiring knowledge of limited aperature and point versus extended source. Thus, known biological damage threshold values for output characteristics similar to the laser being evaluated should be used for confirming practical controls for CW lasers.

Adherence to the ANSI standard classification tables will not, in all cases, guarantee protection, even if the explicit control measures are instituted for those pulsed lasers listed as Class 3b or 4. MPE levels should be considered, but supplementary data is required for unlisted ultra-fast pulses.

Signs on doors notify all entrants to a laser facility that a laser above Class 1 is housed within. The sign required by the ANSI Z136.1 standard should list the properties of the laser, especially the wavelength, as well as the maximum accessible power density or energy density, in addition to entry instructions (see Fig. 5.5).

ENGINEERING CONTROLS

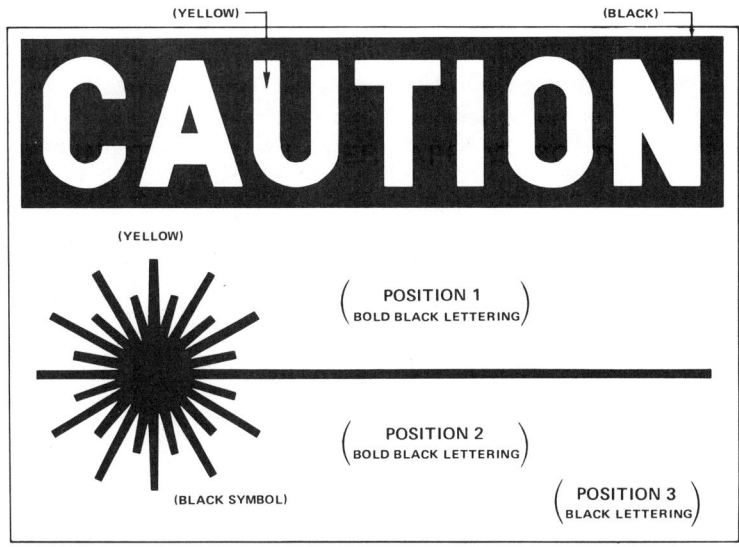

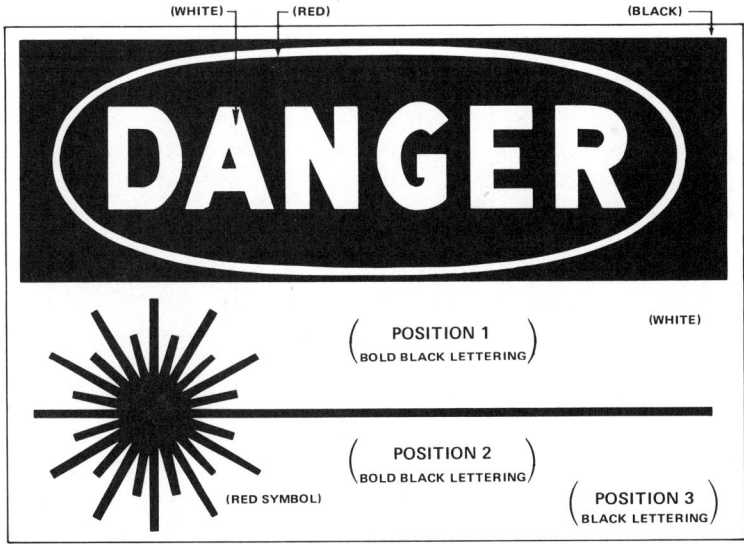

Figure 5.5 Signs required on entry to Class 2 (upper) and for Class 3 and 4 (lower) lasers.

Lights at the entry should warn entrants of the status of Class 3 or 4 lasers, with appropriate wording to explain the entry procedures. Three-light devices have been adopted as a universal warning system similar to traffic light: green (*no restriction on entry*); yellow (*low hazard mode, knock for entry*); and red (*do not enter*). Commercial models are available at reasonable cost.

Door interlocks are devised to protect personnel entering a controlled location for Class 3 or 4 lasers. A variety of schemes is permissible. The standard calls for a design "to prevent firing of the laser" if it is pulsed, and, for Class 3*b* or Class 4 lasers, "the interlocks shall turn off the power supply to interrupt the beam." However, responsibility is assigned to the Laser Safety Officer "for establishment of surveillance of appropriate control measures and to avoid needless duplication in those instances where several alternate yet equally effective control measures may limit exposure." Thus, research laser facilities and clinical medical applications can be controlled by interlocking entry doors to a beam shutter, by an audible alarm, or even by the use of personnel door guards, if approved by the Laser Safety Officer.

Eye protection is recommended for all personnel required to be in laser-controlled areas: (a) if the laser is pulsed, or, (b) if the maximum accessible power density of a CW laser is above 50 mW/cm^2 for wavelengths in the ocular region, and above 500 mW/cm^2 for other wavelengths. (Beam diameter determines diffraction-limited spot size. See Chap. 3, Beam Focusing.) These values are policy at Los Alamos and are consistent with known damage threshold values, assuming that no intentional direct viewing is permitted. These values are a simplified approach to practical application of the ANSI standard.

The type of eyewear recommended depends on the wavelength involved and the desires of the wearer. Laser-protective eyewear has been developed (5) by Fred Reed Optical Company, Albuquerque, NM, that provides comfortable, lightweight spectacles and includes lenses of specially colored glass filters capable of providing correction, even bifocals, if desired. The hardened glass provides mechanical protection and is scratch resistant. A variety of safety frames also is available. This type of eyewear can be carried in a pocket case and is immediately available (see Figs. 5.6 and 5.7).

In addition to spectacles with glass lenses—preferred by most laser personnel working routinely with lasers—a variety of plastic goggles and spectacle styles is available for special purposes (i.e., visitors, short-term employment, short-term laser use), but glass is the preferred absorbing media because (1) prescription lenses are available, (2) thickness can be varied, (3) visual transmission is superior, and (4) damage threshold values are much higher.

Tradeoffs usually are required in the selection process weighing the various options in style, filterability, visual transmission, comfort, durability, and, of course, cost effectiveness (some plastics may have a tendency to scratch easily and thus are not as cost effective as hardened glass, if long-term use is anticipated).

INDOCTRINATION AND TRAINING

Figure 5.6 Examples of glass lenses in spectacle frames available for laser beam protection. Demonstrator is wearing Schott Optical Glass Company's laser glass, designated KG-3, a clear glass that absorbs infrared wavelengths starting at 1.0 μm. The darker eyewear, BG-18, absorbs the red wavelengths of the krypton laser used for protection during alignment procedures at the Antares CO_2 project at Los Alamos. (Courtesy of Los Alamos National Laboratory, Los Alamos, New Mexico.)

Usually, if the options are described to the laser user, chances are better that the eyewear selected will be worn—the most important objective of providing eye protection. Spectacle styles of eyewear are shown in Figures 5.6 and 5.7. Figure 5.8 lists glass and plastic eyewear from Fred Reed Optical Company, a full spectrum eyewear supplier. Practical optical densities are suggested.

A 16-mm color sound motion picture (6) is available for laser eyewear indoctrination.

INDOCTRINATION AND TRAINING

The laser user should be aware of the potential hazards posed by a laser before operating it. The extent of knowledge required for safe operation should be

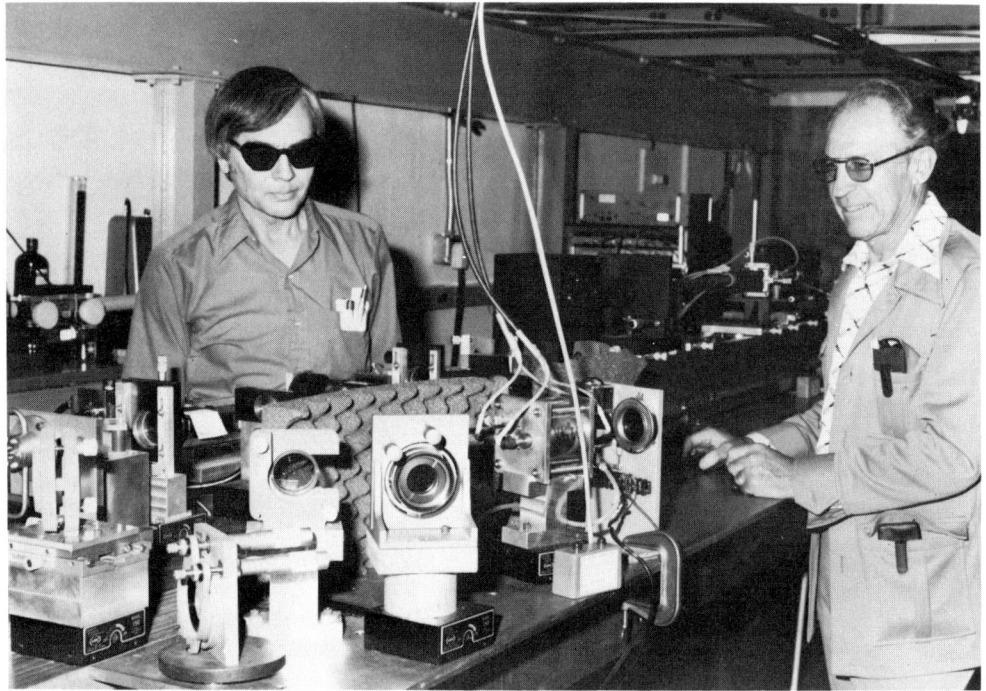

Figure 5.7 Same eyewear as shown in Figure 5.6 but demonstrated in a Nd:glass laboratory. Both glasses absorb the 0.6 m wavelength. In addition, the BG-18 lenses on the left were also used in this facility when a diagnostic ruby 0.69 μm) was in use. (Courtesy of Los Alamos National Laboratory, Los Alamos, New Mexico.)

determined before in-depth instructions are presented to nontechnical users on the fundamentals of laser physics, biological damage thresholds, optical density calculations, etc. However, a brief indoctrination of all laser personnel by the Laser Safety Officer should be offered, not only to familiarize laser users with the rudiments of laser hazards control, but also to present policies of the employer. It also is an opportunity to discuss associated hazards of laser operation; electrical safety should be given special attention. A manual (7) on the Los Alamos laser safety program describes policies for control of laser and associated hazards.

Routine, repetitive operation of a laser may require only specific training for that particular laser, and the operator then may be "qualified" or "certified," as is often done in industrial applications.

Research and development programs that include a variety of laser applications should include a more extensive indoctrination of personnel. Often the laser unit or system is modified or changed, and the hazards of the resulting oper-

ADMINISTERING A LASER SAFETY PROGRAM

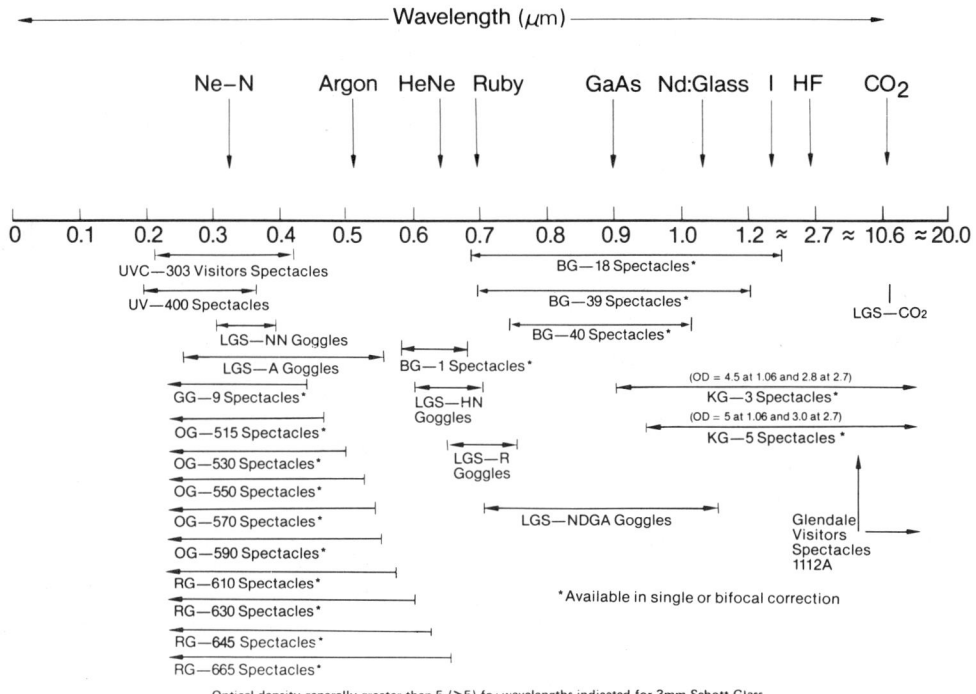

Figure 5.8 Selection chart for locating options of eyewear for various laser wavelengths. (Courtesy Fred Reed Optical Co.)

ations must be reevaluated by knowledgeable personnel. Often, if the program is extensive, one of the users is assigned to be the safety specialist and, with encouragement from the Laser Safety Officer, may attend a one- or two-day course.

Laser safety officers themselves may require special training. Often the responsibility for laser hazards control is assigned to a safety engineer without previous exposure to laser safety information. Extensive indoctrination may be considered. Attendance at a short course may be appropriate as shown in Figure 5.9, in addition to a thorough knowledge of the requirements of the ANSI Z136.1 standard. It is anticipated that this standard will become federal law in the near future.

ADMINISTERING A LASER SAFETY PROGRAM

Constant surveillance is required for the administration of a successful safety program. Routine, regular, documented inspections of the facilities each quarter

Figure 5.9 Short-course indoctrination for laser safety is recommended for laser personnel, especially laser safety officers.

are recommended. Distribution of the inspection report to all employees of the organization involved, as well as to supervisors and department heads, keeps the subject of safety in the forefront during all hazardous operations. Selecting key personnel to serve on the inspection team is important, especially to help in detecting unsafe conditions and in determining assigned responsibility.

Safety committees are vital in developing policy and recommendations for detailed operation of the safety program. Topics for safety meetings, pertinent details of hazardous operations, and other ideas emanate from inspired committee members.

For large organizations, two levels of committees may be desirable, one composed of members of the working group and the other (higher level) of representatives from each individual group. The latter committee could set policy, determine the effectiveness of the program, and recommend emphasis in current inspections. The lower committee should perform the inspections, recommend action, and conduct meetings involving some applicable aspect of laser safety.

Safety inspections are required to determine compliance with established law and standards. A routine schedule should be set, perhaps quarterly. Conditions that do not conform to policy or safe operating procedures are recorded and reported to those responsible for correction. The inspection team should consist of the Laser Safety Officer and representatives of the group who are knowledgeable about the operation of the facility. Larger organizations may assign safety specialists and maintenance personnel to aid in the task.

Safety meetings should be held routinely (i.e., quarterly), to emphasize a topic germane to the operation of lasers, to inform employees of recent changes in safety rules, to report results of an inspection, or to exchange information on hazard control. Attendance at the meetings should be mandatory. Supervisors, staff members, and group leaders also are expected to attend to show management support of the safety program.

Standard operating procedures are important in providing control of hazards involving Class $3b$ or Class 4 lasers and are required by ANSI Z136. A simple written approach to defining the hazards, the countermeasures employed for their control, and assignment of responsibility is a constant reminder of the proper procedures to be invoked when the laser is in use. If an operation is complex and several people are involved in the operation controls, detailed, step-wise listing of each operating procedure may be advisable. Updating once each year is recommended. Posting a copy at the controls is mandatory.

Medical surveillance of laser personnel need not be an extensive nor expensive program. The degree of exposure to each classification of personnel should be known. The ANSI standard for laser users requires only that "incidental personnel shall have an eye examination for visual acuity." Also, that "laser personnel shall be subject to the following: a medical history; visual acuity; examination of various structures of the eye depending on wavelength." An "appendix" is presented in the standard for details, but it should be pointed out that expensive fundus photographs are not required, nor are repeated examinations. If only wavelengths outside the ocular focus region (0.4 to 1.4 μm) are involved, retinal examinations are not required, so only external components need examining.

Consultants in laser safety may be employed to analyze laser applications and to give advice on various aspects of the hazards involved. In many instances, the cost is offset by savings in procurement of engineering controls and eyewear, in recommended eye examinations, etc., and by expert, simplified interpretation of the standards and laws.

SUMMARY OF PRACTICAL LASER SAFETY PROGRAM

Planning to use a laser requires an analysis of the proposed laser environment before entering into an approach to control the hazards associated with the operation of the laser. Simple, yet effective, measures often fulfill laser safety requirements. However, knowledge of the laser's characteristics must be available to determine the accessible power density or energy density of the beam. Comparing these values with established biological damage thresholds and using ANSI Z136.1 (1986) Maximum Permissible Exposures data as a guide will help determine the level of potential hazard involved. Because specular (mirror-like) reflection of a beam must be considered, the establishment of controls for a laser beam of intensity below that established by ANSI A136.1 may be desirable.

Laser safety information is available from several sources, including a book on practical laser safety containing a simple digest of ANSI Z136.1 (8), so that proper indoctrination is available at various levels of interest for laser users. Familiarization with ANSI Z136.1 is critical, because a revised version (1986) has recently been published and will become law in the near future. Services of an expert on laser safety often are warranted, especially for nonexperienced entrants into this versatile technology.

REFERENCES

1. American National Standard Z136.1-(1986). *Safe Use of Lasers*. American National Standards Institute, New York.
2. *Laser Classification Guide* (1976). National Institute for Occupational Safety and Health, Department of Health and Welfare, Cincinnati, OH.
3. Winburn, D.C. (1977). Biological Damage Threshold Studies for Fast-Pulse Lasers at Los Alamos. *Electro-Optical Systems Design*, 9 (11), 19-22.
4. Decker, C.D. (1977). Comment. *Laser Focus*, 13 (8), 6.
5. Winburn, D.C. (1979). Laser Safety Eyewear Update. *Electro-Optical Systems Design*, 11 (1).
6. Barnett, C.T., and D.C. Winburn (1979). *Lasers and Your Eyes* (Film No. 328 —14 minutes). Los Alamos National Laboratory, Los Alamos, NM.
7. Winburn, D.C. *Safety Manual for Laser Research and Technology*. Report No. LA-5305-MS, Rev. 2, LANL. Los Alamos National Laboratory, Los Alamos, NM.
8. Winburn, D.C. (1985). *Practical Laser Safety*. Marcel Dekker, New York.

6
Engineering and Science Applications

In the early 1960s, the laser was a laboratory curiosity, an invention waiting to be used. The news media took delight in covering demonstrations showing how this new "toy" could be focused to burn a hole in a hardened steel razor blade, how a laser beam could penetrate a colorless balloon yet cause an interior colored balloon to burst, and various other tricks of physics this wondrous light source could perform. It did not take long, however, before the medical, construction, manufacturing, electronic, and military industries were finding useful applications for the laser. By the mid-1960s, laser manufacturers set up shop all over the country. Some failed, others flourished. But the varieties and types of lasers that emerged from the research and development laboratories found applications in hundreds of ways. It is not intended in this limited work to describe as many applications as possible but rather to select those areas of laser employment that demonstrate potential interest to engineers in general and to provide references to pursue for technical detail. No special order of importance has been assigned to these applications described in Chapters 6-9.

Details on manufacturers of lasers and laser types are contained in an annual publication of Penn Well Publishing Company's magazine, *Laser Focus/Electro-Optics*. This "buyers guide" (1) reports not only the manufacturers but also all essential characteristics of the lasers and accessory apparatus. Other sources of laser suppliers are the publications *Lasers and Applications* (2) and *Photonics Spectra* (3), both monthly magazines of the laser industry.

MATERIALS PROCESSING

To metallurgical and mechanical engineers, a knowledge of laser technology relating to metal working techniques in current manufacturing industries is a requisite. Lasers can cut, drill, weld, remove metal from surfaces, heat treat localized surfaces, alloy metal surfaces, and perform these operations even at surfaces inaccessible by mechanical methods. It is not necessary that users of lasers for these operations understand all the technical details of the laser system, but knowledge of the beam characteristics and material absorption qualities is essential in proper adaptation of the beam to obtain the desired result. For example, a decision on the wavelength must first be made when considering interaction with a specific

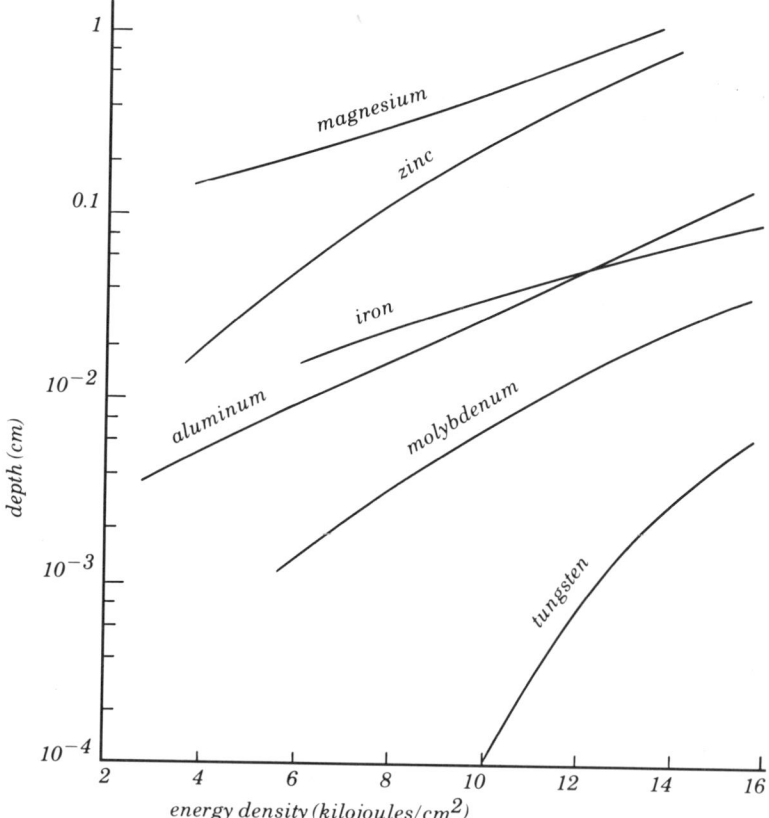

Figure 6.1 Depth of material vaporized vs. energy density curves for various metals irradiated with 700-μs pulse of Nd:glass laser light of 1.06 μm wavelength. (Courtesy of Academic Press, Inc.)

MATERIALS PROCESSING

metal, then a determination must be made as to whether a continuous beam or a pulsed beam would be more efficient. Many of these details have been developed for many metals, and references are listed at the end of this chapter. For example, Figure 6.1 shows the depth of material vaporized in several metals as a function of energy density in a 700-μs pulse of a Nd:glass laser, as reported by Ready (4). The curves for the metals with a low latent heat of vaporization appear above those with a higher heat of vaporization, so that the energy density requirements for a specific application for one of the metals shown can be selected if the Nd:glass pulsed laser is to be employed.

Figure 6.2 CO_2 laser capable of cutting, welding, or heat treating a wide range of materials. (Courtesy of Coherent, Inc., Palo Alto, California.)

Cutting and Drilling

Because CO_2 lasers, which produce 10.6 μm wavelength beams, can be manufactured with extremely high continuous wave (CW) power densities, this laser system (Fig. 6.2) is the most commonly used in industrial cutting and drilling operations not only with metals but also such nonmetals as ceramics, plastics, cloth, paper, glass, and so on. Gas jetting is usually associated with industrial metal cutting either to enhance removal with a reactive gas, such as oxygen, or with a protective (inert) gas to protect flammables. Typical metals that are efficiently cut or drilled with the CO_2 systems are steel, aluminum, and titanium. Cutting speeds reach rates from 20 inches per minute for one-inch carbon steel sheets to 100 inches per minute for one-half-inch aluminum alloys (5). Laser energy supplies all the energy needed to cut or drill nonmetals, but often coaxial gas jets serve to protect the treated surfaces and coaxial vacuum can be included to direct toxic fumes away from the operation.

Welding

Several advantages over gas or arc welding are possible with laser-welding techniques; (a) purity of the materials involved is not altered, (b) localized heating by the small spot size can be accurately controlled, even programmed, by computers to reproduce exact characteristics, and (c) location on assembly line applications (see Fig. 6.3), and the laser beam can be transmitted through windows of a closed container to permit welding (or other operations) in a controlled atmosphere. For example, one application at the Los Alamos National Laboratory required sampling of a gas contained in a sealed metal container within a "hot" cell. The solution: a ruby laser beam penetrated the leaded glass window of the cell, was focused on the metal container within a glass vacuum system and a hole cut to release the accumulated gas for analysis. Laser welding also permits welding of dissimilar materials and complex geometries (using mirrors and lens systems). Interfacing with robotics is another advantage (Fig. 6.4).

Heat Treatment

Although induction heating has served the metal processing operations for many years in providing surface treatments, for hardening steel particularly, the versatility of the laser beam has proven very useful in the treatment of many metal surfaces. Cost savings is one consideration, because less energy is required, heat losses are reduced, and the time involved is considerably lessened. Steel and aluminum surfaces can be hardened or alloyed by methods established for production line applications. A less expensive substrate can be used in some applications so that metal powder layers can be coated and then processed thermally by the laser beam in precise locations.

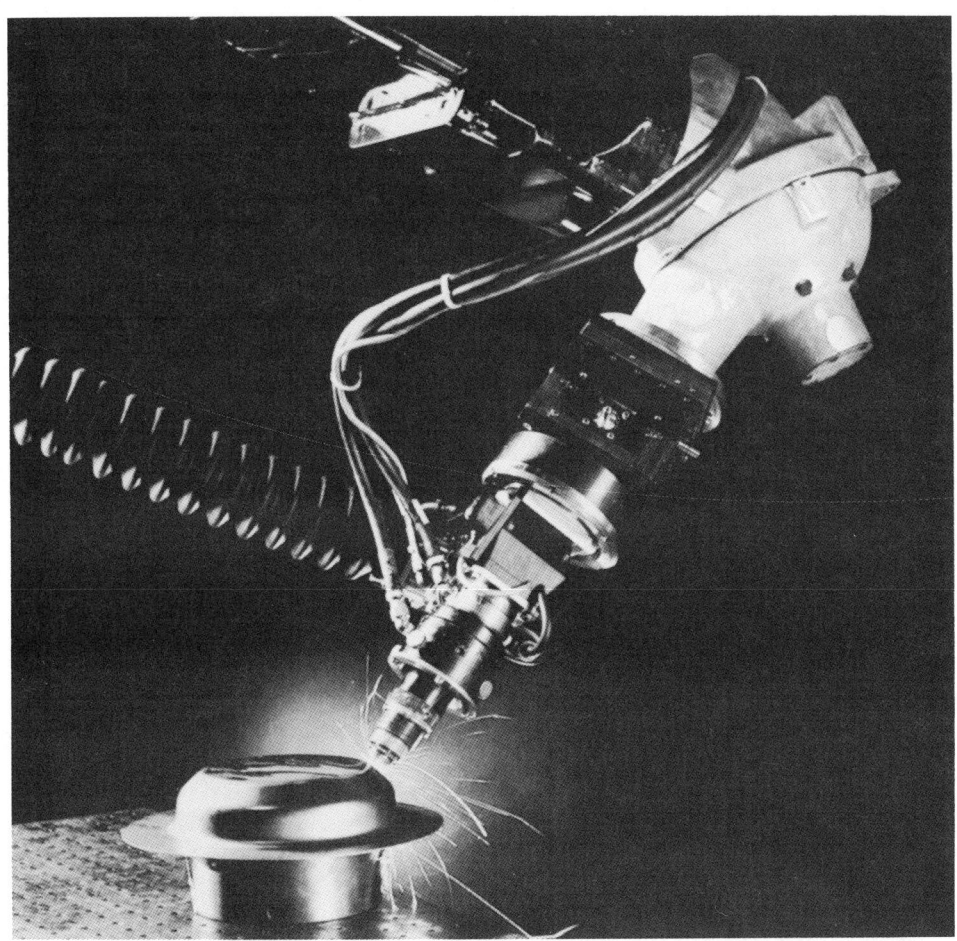

Figure 6.3 A Cincinnati Milacron 776 robot coupled to a Spectra-Physics Laserflex beam delivery system demonstrates robotic manipulation of a laser beam in three dimensions for materials processing. The laser is a 1500-W industrial CO_2 laser. (Courtesy of Spectra Physics.)

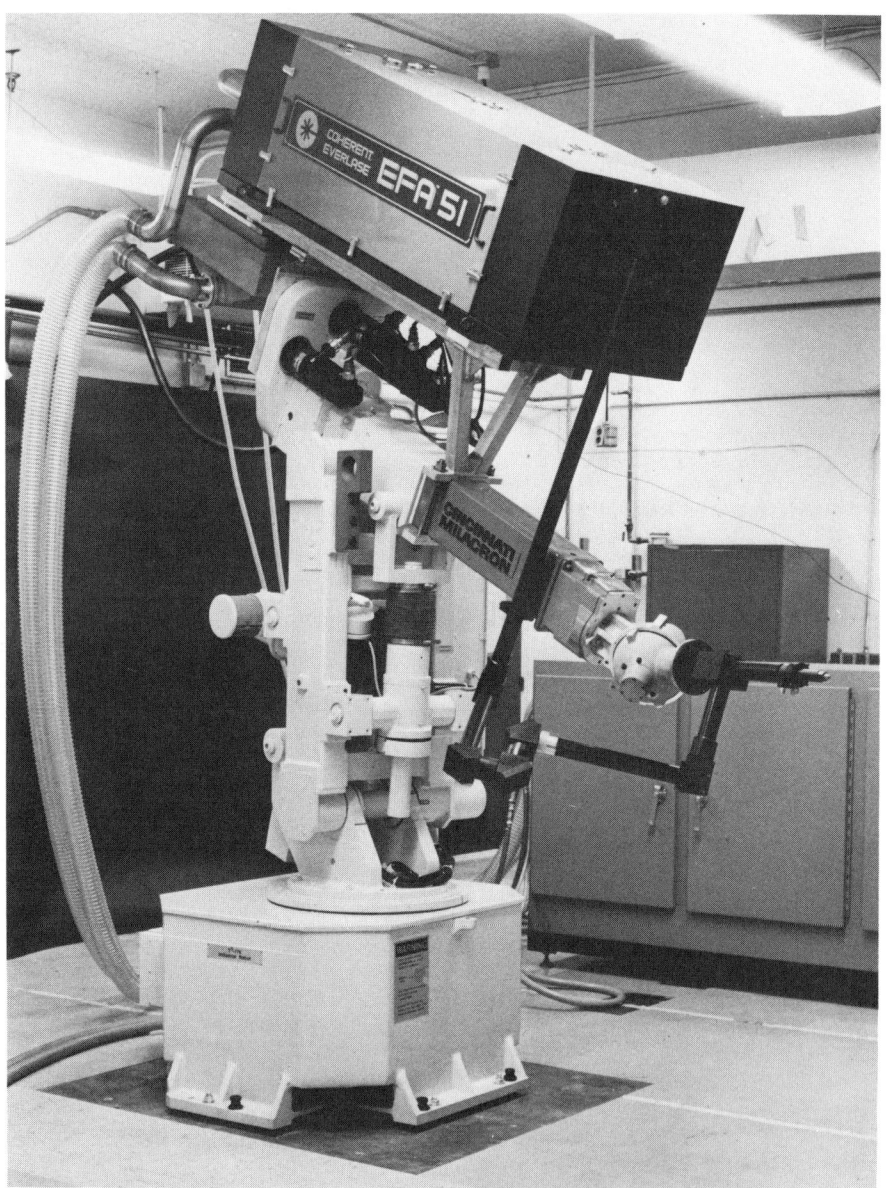

Figure 6.4 Integrating a high power CO_2 laser with industrial robot for materials processing. (Courtesy of Coherent, Inc., Palo Alto, California.)

Semiconductor Processing

According to an article in the May 1986 issue of *Lasers and Applications* (2), excimer lasers are extending the capabilities of silicon and gallium arsenide processing techniques, because current trends in the semiconductor industry place increasingly more stringent demands on chip fabrication techniques. Higher chip densities demand smaller feature sizes; dimensions of 0.5 μm should be commonplace before 1990. Furthermore, shallower depth profiles are required for modern integrated circuits. Junction depths on the order of 0.1 μm are needed for submicrometer metal-oxide semiconductor technology.

Another factor to be reckoned with is chip customization. Over the next several years, most integrated circuit (IC) production will switch from the large production volume stock configurations common today to more specialized, smaller volume chip types common to a broader range of microprocessor-controlled products.

Simultaneously, wafers are becoming larger, while economic considerations demand higher throughputs and improved yields.

Many established methods of semiconductor processing are thus becoming steadily less suitable for current needs and will likely be even more out of place soon. Considerable effort in developing new and often revolutionary production techniques is being expended in many laboratories today.

Excimer lasers will have a significant role in this development. Excimers can greatly increase the capabilities of conventional photobased processes such as lithography, which currently employ lamps as optical sources. Other entirely new excimer laser-based techniques in areas such as film deposition and removal are not possible with lamps. Thus, some potential applications areas use the excimer laser simply as an improved lamp, while others make particular use of its laser properties. The distinction is often not a sharp one.

The principal attribute of excimer lasers for semiconductor processing is the availability of high average powers in the ultraviolet. This wavelength region offers improved resolution, the potential for photochemical rather than thermal control of reactions, and a means of restricting surface reactions to very shallow depths. Before considering these factors in more detail, it is worthwhile to review existing sources of deep ultraviolet radiation.

The most commonly used light source in this wavelength region is the discharge lamp. The output power of a lamp falls off sharply in the ultraviolet, which is precisely where excimers operate. Furthermore, although an excimer laser efficiency of a few percent may not seem high compared to a lamp's electrical-to-optical conversion efficiency of 50% or more, all the radiant output of the excimer laser is usable power. For a lamp, most of the optical output is in the visible and infrared regions of the spectrum and must be filtered out lest it damage the wafer or produce undesired side effects in the production process.

The inefficiencies associated with excimer lasers are such that the input power not converted into optical output is dissipated within the laser itself, unlike lamps which require output filtering.

Lasers have long been recognized as an alternative to lamps, nor are excimer lasers the first ultraviolet lasers to be used in semiconductor processing. Ion lasers, frequency-doubled ion lasers, and fourth-harmonic Nd:YAG lasers have been and are still used, although they generally produce output powers of less than 1 W. Moreover, the utility of such lasers in any imaging application has often been limited by the occurrence of speckle.

Speckle is an interference effect arising from the high degree of spatial coherence of such lasers, many of which operate in a single transverse mode. The problem with speckle is that it leads to severe nonuniformities in exposure. Conversely, free-running excimer lasers, with their large area beams and short cavity lifetimes, have relatively poor spatial coherence. Consequently, an excimer laser's output is highly multimode (as many as 10^5 transverse modes). Experimentally, speckle has not proved to be a problem when excimer lasers are used in imaging applications.

In short, the high output powers available in the deep ultraviolet and the absence of coherent effects have made the excimer laser an attractive light source for semiconductor processing. The applications of excimer lasers to semiconductor processing can be grouped into three main areas: photochemically driven processing, surface processing, and applications dealing with the excimer's increased resolution capabilities.

One major advantage of excimer lasers is the high photon energy (up to 7.9 eV) associated with the short wavelengths. Because this energy is greater than the binding energies of many molecules, at sufficiently high fluence it is possible to photodissociate such molecules and hence initiate or control photochemical reactions. In many cases, this represents a very attractive alternative to conventional processing techniques, in which high temperatures or plasma discharges produce the desired reactions. Excimer-laser-assisted film deposition and removal are two examples of this capability.

In conventional thermally driven chemical vapor deposition (CVD), substrates and reaction chambers are often heated to temperatures in excess of 600°C. Unfortunately, these high temperatures can produce undesirable side effects including uncontrolled diffusion of dopants, wafer warpage, and defect generation and propagation. These issues were of only minor concern in the era of small wafers and coarse feature sizes, but they can drastically reduce yields in present-day chip fabrication.

A significant improvement in this situation was the advent of plasma-controlled processing. Here, a plasma is used to create the reactant species, thereby allowing a substantial decrease in processing temperatures and the associated ill effects. Unfortunately, plasmas bring a new set of problems: radiation damage of films and substrates (particularly for gallium arsenide) and increased contaminants from ion bombardment of the reactor vessel. Above all, the plasma processes are complex and allow many competing reactions to occur, while the narrow operating window in which a stable discharge can be maintained limits the versatility of this approach.

Excimer lasers offer a straightforward alternative to these traditional techniques. In excimer laser CVD, the laser beam enters a cell containing the substrate and a donor gas, both of which are at or near ambient temperature. The donor gas absorbs the laser radiation and dissociates, producing the reactant species which results in the deposition of the desired film.

In the case of metal films, the deposition can follow directly from dissociation of donor gases such as trimethyl aluminum or tungsten hexafluoride. Using this technique, George Collins and his group at Colorado State University have achieved deposition rates of 250 nanometers per minute, even over relatively large areas. The films showed high adhesion, good uniformity, low mechanical stress, and resistivities within a factor of two of bulk values. Also, irradiation of the sheet of gas directly above the substrate results in a planar source of reactant material and hence excellent step coverage, a distinct advantage over evaporation or sputtering techniques. Metals deposited in this manner include molybdenum, tungsten, chromium, aluminum, zinc, and indium.

The Colorado State group also produced similar results in depositing insulating films such as silicon dioxide and aluminum oxide. For these films the excimer laser dissociates one donor species, which then reacts with a second donor species to produce the compound film. The fact that the argon-fluoride laser radiation at 193 nm is only weakly absorbed by the silane illustrates another advantage of using the excimer laser as the light source: the laser's monochromaticity guarantees driving only specific predetermined chemical reactions. The excimer laser deposition technique has also been extended to semiconductors such as polycrystalline silicon and III-V materials such as indium phosphide.

In a variation of the CVD scheme, a group at Fujitsu Laboratories demonstrated excimer-laser-enhanced nitridation of silicon substrates. Specifically, the beam from an ArF laser-dissociated ammonia gas, and the resulting NH_2 radicals reacted with the silicon substrate to form silicon nitride. Although in this case the substrate was irradiated at near-normal incidence, the induced temperature rise was negligible, verifying the photochemical nature of the process.

This process, analogous to laser deposition, is a viable alternative to conventional etching techniques. Historically, wet etching was used in most applications. However, wet etching introduces high levels of contamination, so dry (i.e., plasma) etching techniques have gained prominence when reduced feature sizes are involved. Unfortunately, plasma etching techniques suffer from the same drawbacks as plasma-enhanced CVD.

Excimer lasers again offer an attractive alternative. As in laser CVD, the substrate is contained in a reactor chamber filled with donor gas, which is irradiated by the excimer laser. The resultant dissociation produces large concentrations of radical species, leading to a rapid etch rate.

Another film removal process is that of direct laser etching. Here, the laser acts directly on the substrate without the assistance of any donor gas species. A classic example is photoablative decomposition, which occurs when organic materials are irradiated with excimer radiation or sufficient fluence.

Initially, researchers proposed that the process was entirely photochemical, with the excimer laser pulse breaking the bonds of the substrate material and the resulting fragments ejected from the irradiated area carrying off the excess energy. More recently, others have suggested that a thermal component is also involved.

Nonetheless, a sharp energy fluence threshold exists. The surrounding material is largely unaffected and free from charring, in sharp contrast to the burning action often associated with other laser-based material-removal techniques. This process has led to the discovery of self-developing resists such as nitrocellulose films that volatize upon exposure, eliminating the need for any subsequent development steps. Metal film removal by direct excimer irradiation has also been demonstrated. In this case, however, the mechanism is simple laser-induced evaporation caused by the rapid heating of the surface layers, as opposed to any photochemical process. The ultraviolet radiation is strongly absorbed by most metals, and the short pulse duration minimizes the heat-affected zone.

Although various trends in the semiconductor industry mentioned at the outset of the chapter have spurred considerable process development, nowhere is this more evident than in lithography. Dramatic improvements in the more conventional optical lithographic techniques have paralleled the development of revolutionary approaches based on electron beams and x-rays.

Once touted as the obvious successor to photolithography, e-beam lithography has proven too expensive and of too low throughput to be of much interest, except for specialized wafer fabrication requirements. X-rays offer the potential for almost limitless resolution at reasonable cost effectiveness, but considerable work on x-ray sources, masks, and resists will be needed before this technique is ready for large-scale use.

These factors, together with the evolutionary improvements in photolithographic exposure systems and resists, have maintained the popularity of optical lithography in today's semiconductor industry. Over the past several years, investigations into the use of excimer lasers have extended the capabilities of optical lithography even further.

New photoresists and improved registration techniques must be developed to realize the full potential of excimer lithography. But these changes are minor compared to those required, for instance, to switch to x-ray lithography.

Jain and co-workers at IBM first demonstrated excimer photolithography in 1982 and reported his findings in a published article. By using a variety of excimer laser wavelengths and resists, they produced submicrometer feature sizes with exceptional clarity. Not only was there no evidence of speckle, but standingwave effects, a concern when using a monochromatic light source, were barely noticeable. Moreover, they showed that reciprocity failure did not occur.

To probe the limits of excimer photolithography, Harold Craighead and his co-workers at Bell Laboratories used the shortest wavelength excimer, a fluorine laser emitting at 157 nm, to achieve resolutions better than 0.2 μm. Although the F_2 laser is unlikely ever to be a practical source, these experiments serve as an indication of the ultimate capabilities of photolithography.

The true value of excimer lasers in lithography does not lie in their short wavelengths, but rather in that they possess considerable output power at these wavelengths, typically several tens of watts of average power. In contrast, good discharge lamps generally emit only several tens of milliwatts in the deep ultraviolet. Even after the requisite beam handling in each case, the improvement in exposure time is dramatic. As an example, for a commercially available contact printer, the exposure time for a four-inch wafer using an ArF laser is two to three minutes, compared to twenty minutes using an ultraviolet lamp on the same system.

Another potential mode of operation offered by excimer lasers is that of "flash-on-the-fly" lithography. With conventional lithographic steppers, each optical field on a wafer is brought into alignment with the exposure system, the wafer is held stationary, and the exposure is made. This process is repeated until the entire wafer has been covered. Excimer lasers possess sufficient output energy to achieve the desired exposure in a single pulse. Moreover, the excimer laser's short pulse duration represents an instantaneous dose of optical power, at least on the time scale of mechanical movements. Thus, the notion of flash-on-the-fly becomes possible. Here, the stepper would no longer need to step in the sense that the wafer would have to stop at each field during the exposure. Instead, the total exposure can be achieved in the very instant that the moving wafer is in proper alignment. The resulting improvement in throughput would be a factor of two or three, making this technique an extremely attractive possibility.

Increased resolution can also be achieved by direct excimer laser writing. As with imaging, the short wavelength of the excimer laser allows direct focusing of the beam to micrometer-size spots. The resulting spot could then be translated across the substrate surface. This, in turn, introduces some interesting possibilities for direct "writing," "erasing," or, more specifically, localized film deposition or film removal. These are simply the direct-write analogs of the broad-area techniques discussed previously.

Increasing chip customization, where a family of chips could be fabricated from a basic stock starter chip, has intensified interest in this excimer processing capability. Such processing could be accomplished by making or breaking specific interconnections between various regions on a chip. Such an approach has already been demonstrated in the case of customized gate arrays, using an argon-ion laser as the light source. The extension to excimer lasers appears straightforward, generating the improvement in resolution offered by the shorter operating wavelength. Limitations of the pulse repetition rate on scanning speed would need to be considered, however.

In semiconductors, like many other materials, optical absorption tends to increase with decreasing wavelength. In silicon, the absorption depth decreases from several tens of micrometers in the infrared to perhaps 1 μm in the visible and to approximately 0.1 μm in the ultraviolet. Consequently, irradiation of materials such as silicon by an excimer laser causes a substantial deposition of energy very near the surface. This, in turn, can lead to melting of the surface layers. But the dramatic aspect is that the melted silicon regrows as undamaged single crystalline material. This has led to the development of several processing techniques, most notably annealing and doping.

Until recently, most semiconductor annealing was performed in furnaces. However, as feature sizes began to shrink, dopant migration and wafer distortion stemming from prolonged high temperature conditions limited the use of furnace annealing for many applications. This situation led to the development of transient heating techniques, most notably rapid thermal annealing (RTA) and pulsed laser annealing.

In RTA, the wafer is heated by high-intensity lamps, reducing the high temperature phase from tens of minutes, as required in furnace annealing, to tens of seconds. Processing time can be reduced much further by using pulsed laser annealing techniques. Pulsed laser annealing techniques were first demonstrated in solar cell fabrication using solid-state lasers. More recently, using excimer lasers, largely on account of higher optical beam quality, superior results have been produced.

Excimer laser annealing applied to IC fabrication is not as well developed as it is for solar cells, but preliminary results are encouraging. Work performed at Hughes Research Laboratories on the annealing of silicon-on-sapphire structures provides an example. By exposing these structures to KrF radiation, several beneficial effects can be realized: improved edge contours, reduced crystal defects, and better step coverage in subsequent metallization. Metal-oxide semiconductor (MOS) transistors fabricated over these laser-annealed structures showed a 30% increase in channel mobility.

When ions are implanted to dope the semiconductor material, crystal lattice damage occurs. Soon after the development of laser annealing techniques, practitioners realized that the doping and annealing steps could be combined if the laser-melting step were performed in the presence of the appropriate donor gas. This eliminates the need for the ion-implantation step altogether.

With the excimer laser, in particular, melting is restricted to a volume within a fraction of a micrometer of the surface, thus allowing fabrication of very shallow doping profiles. This technique was demonstrated by researchers at Stanford Electronics Laboratory, who made p-type diodes with junction depths of 0.08 to 0.16 μm and excellent electrical characteristics. In this work, the wafer was placed in an atmosphere of diborane gas and irradiated with an XeCl laser. The resulting profiles were not only shallow but also did not display the deeply penetrating tail associated with ion implantation techniques.

Excimer laser heating can also be used to control chemical reactions at the substrate surface. In this manner, researchers at Xerox have fabricated SiO_2 films on silicon substrates with growth rates similar to those obtained with the laser-deposition techniques discussed earlier. This straightforward technique involved using an XeCl laser to heat the surface of a silicon substrate immersed in an oxygen atmosphere to near (or above) its melting point. The bulk of the wafer remains at its ambient temperature. Because the process is strongly temperature dependent, the transverse oxide profile can be much sharper than the optical beam profile. Thus, this process is well suited for direct-write applications as well.

The article concludes that excimer lasers have an enormous potential in the field of semiconductor processing. Considerable work is still required in process development and on excimer lasers themselves before these processes will be ready for large-scale application. However, the advantages offered by excimer laser processing are substantial, and it appears certain that excimer-based techniques will come to fruition in the not too distant future.

Other Materials Processing Applications

The microelectronics industry employs the laser beam to "trim" resistors in thick film or hybrid circuits, a procedure requiring microscopy for the location and extent of material removal. The inertial confinement fusion, or laser-induced fusion, research programs also use laser microscopy in drilling (5 μm diameter) holes in the microballoons (150 μm diameter) that contain the fusion fuel: isotopes of hydrogen (deuterium and tritium). These uses of laser energy demonstrate the potential of extremely low power applications in materials processing. The ultimate, perhaps, in materials processing application, is research being done at Cornell University with the objective of removing selected atoms from the DNA molecule and studying mutations of cells resulting from selectively removing one or more atoms as desired.

COMMUNICATIONS USING FIBER OPTICS

Lasers play the essential role in using thin strands of glass fibers (see Fig. 6.5) to transmit light signals that can be received and translated into a communication

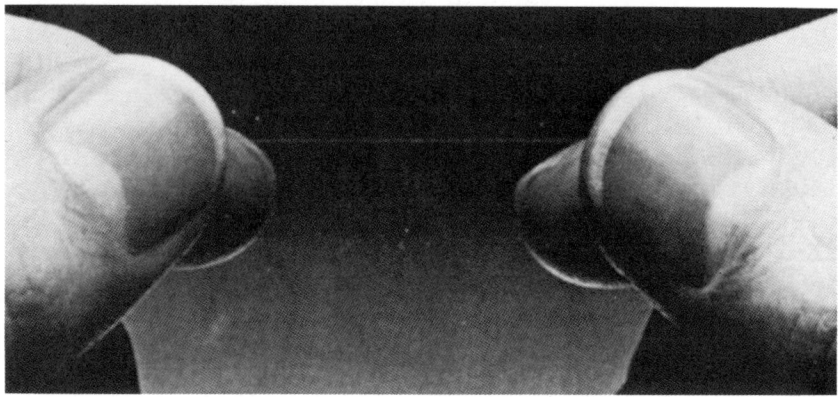

Figure 6.5 Photo showing relative size of fiber used in optical communications.

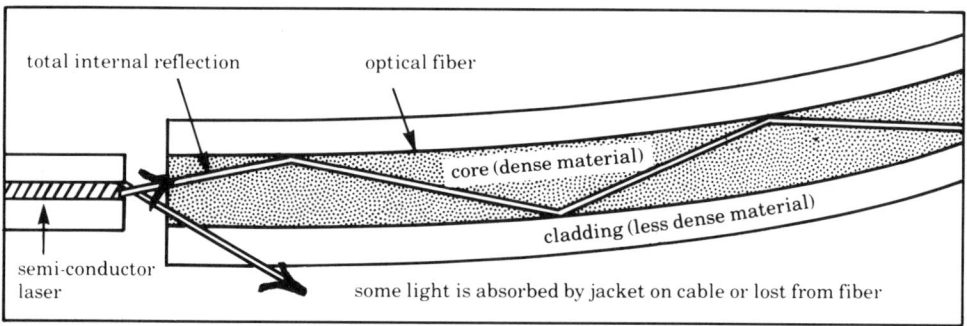

Figure 6.6 Illustrative sketch showing method of transmitting laser beam energy within special glass fiber.

format. The method of transmission is shown in Figure 6.6, where the laser beam is reflected at low angles down the fiber core. Some light is absorbed, of course, or lost, so that the "signal" must be reinforced at some length down the fiber. However, the technology has advanced to the point whereby "repeater" stations are not required for many miles—and the development of improved fibers and research with various solid-state lasers will undoubtedly increase the distance a signal can propagate. These distances are so great now that the need for reinforcing the signal will not restrict installing fiber cables under oceans to provide new communication systems between countries separated by large bodies of water.

The method of signal transmission, which is superimposed on the beam, includes a process called modulation. The light signal is transmitted through the

COMMUNICATIONS USING FIBER OPTICS

fiber to a decoder, or demodulator, that receives the digital photon signal, converts it to an electrical signal, amplifies it, and sends it on for interpretation by a communications network. The contribution which fiberoptics has made in the field of communications and other technologies employing electronic or optical signals has been extensive, from replacing control wiring in automobiles to replacing essentially all telecommunication cables, local and international. Advantages of signals transmitted by photons in fibers versus electrons in metal wires are many. The most obvious are the reduced size and weight, the reduced cost of hardware, and the increased signal capacity. Because the optical system is not electrical, static or "noise," is not encountered. Also, no sparks can be generated so that this aspect of safety in preventing fires or explosions can be utilized in chemical and other industries. Also, light-emitting diodes (LEDs) can be used instead of lasers for short hauls, such as automobiles, computers, printers, etc. (see Fig. 6.7).

The November 1985 issue of *Laser Focus* (1) describes how AT&T plans to test the first undersea fiberoptic cable installation in the Canary Islands in 1986, preceding its transoceanic application in a project called TAT-8. The following has been extracted from that article.

> The Canary Islands test, dubbed the "Short-System Experiment," has two objectives: to confirm the design and to provide the needed high capacity in that popular tourist area. The system connects Tenerife and Grand Canaria, located in the Atlantic Ocean off the northwest coast of Africa.
>
> The prototype undersea cable includes six fibers, four for transmission and two as the spare, and will be 123 km in length with three repeaters when the system is completed. The Spanish telephone company, CNTE, (Compania Telefonica Nacional de Espana) will initially have use of one pair of fibers; the second will be used by AT&T engineers for observation. Eventually, the entire system will be turned over to CNTE.
>
> The undersea cable uses singlemode fiber manufactured by AT&T Technologies, Atlanta, Ga. Continuous-draw lengths of up to 10 km were created using the MCVD process. The fiber was shipped to the Simplex Wire and Cable plant for cabling. Cable lengths were 55-60 km when they were joined together.
>
> The Simplex company has a long-term contract with Bell Laboratories to produce the cable. It has extensive testing capabilities, including a temperature chamber capable of accepting a 10-foot diameter cable reel and exposing it to temperatures from $-20°C$ to $+60°C$. An 80,000 pound tensile test machine can also test a sample of cable up to 100-m long. This machine has a reverse bend device capable of cycling the cable under tension to simulate mechanical loading during installation over the bow sheaves of a cable ship.

112 ENGINEERING AND SCIENCE APPLICATIONS

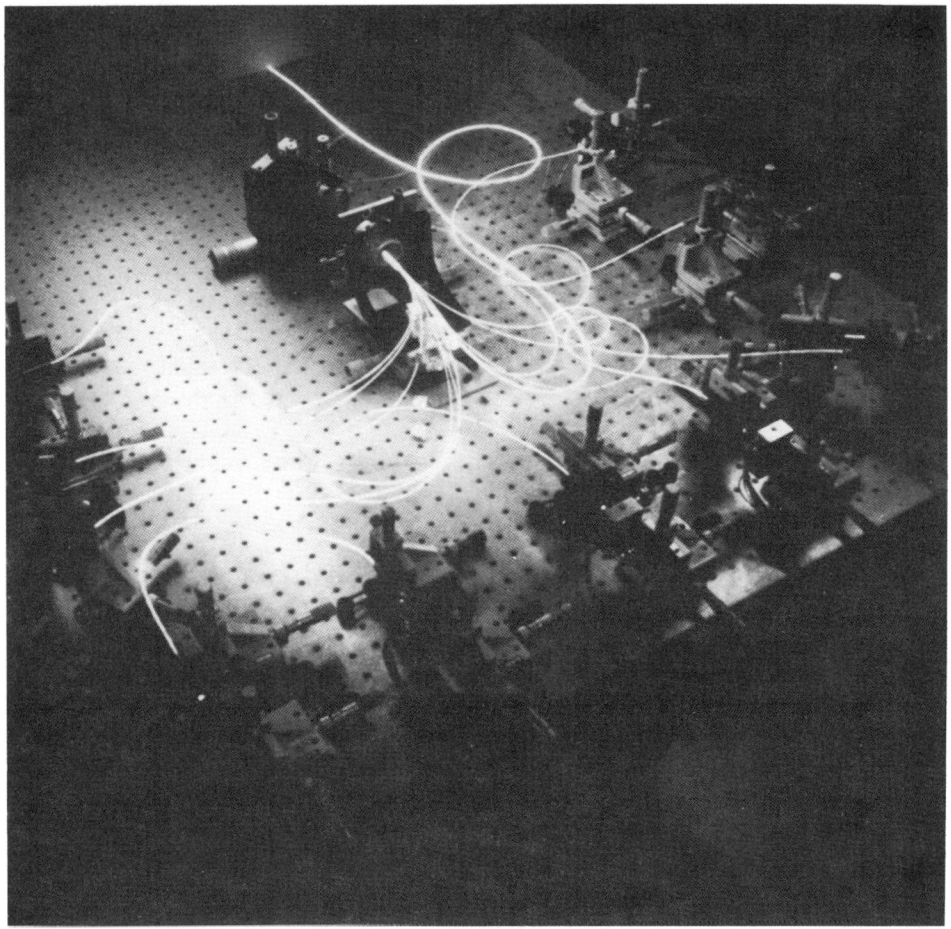

Figure 6.7 Laboratory setup showing an experimental lightwave multiplier combining ten distributed feedback lasers, demonstrating the feasibility of sending 20 gigabits per second through a single optical fiber. (Courtesy of AT&T Bell Labs.)

In addition to the six fibers, the undersea cable contains a central steel wire in its core. Surrounding the core are two layers of steel strands to provide strength and a continuously welded copper jacket to conduct power to the undersea repeaters. A layer of polyethlyene surrounds the copper jacket to provide insulation. The completed cable measures 21 mm (0.8 inches) in diameter.

The system in the Canary Islands was to be installed with two repeaters. A third will be added during a cable repair test. The repeaters are designed to be spaced apart at a maximum of 52 km (30 miles). They were manufactured by AT&T Technologies in Clark, N.J. TAT-8 will have some 125 repeaters.

Each repeater contains circuits for regeneration as well as for monitoring and supervision. The lightwave signal is received at the optical regenerator by an InGaAs *p-i-n* photodiode coupled to a silicon integrated circuit, which converts the signal into an electrical stream. Integrated circuits also perform amplification, timing recovery, and transmitter-driving functions. A buried heterostructure injection laser converts the electrical signals back into optical pulses which are then transmitted over the next cable span. The InGaAs lasers operate at 1.3 μm, and each regenerator contains six spare lasers which can be automatically activated upon command from shore.

The shore terminal includes three major components: a power plant that can provide up to 7500 V to power the undersea repeaters; a supervisory terminal to monitor the regenerators and to control switching; and a transmission terminal to interface between the electrical pulse streams and the optical cable.

This summer marked the 35th anniversary of the first undersea cable test bed. That system ran between Florida and Cuba, and preceeded the installation of the first transAtlantic cable in 1956 (TAT-1). The Bell Telephone Laboratories developed the system, which provided 36 circuits using a twin cable system, one cable for each direction of transmission. By the time the sixth transAtlantic cable was laid in 1976, the service was providing 4000 two-way voice circuits on a single coaxial cable.

TAT-8 dramatically raises capacity; it will provide 40,000 channels. AT&T has invested about $250 million, including the cost of hardware and development, on TAT-8, according to Frank Tuttle, director of international network and customer services for AT&T Communications and general manager of the Transoceanic Cable Ship Company. The entire TAT-8 project is projected to cost $335 million. A consortium of 29 companies and organizations from Europe and North America will own and operate the TAT-8 system.

TAT-8 will run from Tuckerton, N.J., to a branching unit off the edge of the European continental shelf. The branching unit will allow the cable to split into two segments that go ashore in the United Kingdom and in France. Standard Telephones and Cables of the U.K. will build the segment from that branching point to Widemouth, England, and Submarcom of France will provide the segment running to Penmarch, France.

Following the TAT-8 installation, AT&T Communications will install a similar fiberoptic system from California to Hawaii in 1988, and then to a branching unit about 800 miles northeast of Guam and 1000 miles southwest of Japan. That system, planned for service in December 1988, is called Hawaii 4/Transpacific 3. After that system is in place, AT&T's cable ship will install another 1500-mile section from Guam to the Philippines. That system is tentatively set for service in 1989.

There are many other extensive fiberoptic systems in various stages of completion, including communication links in an entire city in Japan, communication lines in populated "corridors" such as cities near the east and west coasts of the United States, Canada's extensive link, cable TV system, and so on, not only in the United States, but throughout the world. Even in New Mexico, a state that has extensive copper mining interests, glass fibers are replacing copper wires in an extensive communications cable connecting high-tech communities along the Rio Grande corridor, including Los Alamos National Laboratory in the north, through Albuquerque's facilities, past Socorro's School of Mines, ending in the New Mexico State University campus at Las Cruces.

Competition in communications by fiberoptics will come from an extensive satellite system, but cost, security, weather, and other considerations preclude satellite competition for the current high rate of expansion of the fiberoptics industry. Society, even in underdeveloped countries, will benefit from this application of laser technology.

The future of fiber communications appears bright according to Linn F. Mollenauer of AT&T Bell Laboratories in his invited talk at the 1985 Optical Society of America Annual Meeting and reported in the May 1986 *Optics News* (6). "In an 'all optical' fiber system—one without electronic repeaters—a single fiber could transmit as much as 100 Gbit/sec over thousands of kilometers. Such performance would be obtained by using optical gain to overcome fiber loss and by transmitting the signals as nonspreading, soliton pulses." According to the article, the soliton is a pulse of the proper shape and critical peak power, such that effects of index nonlinearity exactly cancel dispersive broadening. The soliton laser is described in *Optical Letters 9* and *13* (1985) by Mollenauer and R.H. Stolen (6).

In a conventional communications long-distance fiber optics system, the optical signals are detected and electronically regenerated every 20 to 100 km before continuing along the next span of fiber, and the electronic repeaters limit rates to 1Gbit/sec or less per channel. An all optical signal system would overcome the fiber losses by making the fiber itself a distributed amplifier through the use of the stimulated Raman effect. Technical details of this approach are described by Mollenauer, and reference is made to other investigators who have reported experimental results in the amplification and reshaping of optical solitons in glass fibers.

In concluding the article, the author states that, "The ideas outlined here will soon be given experimental test in a closed loop of fiber whose length will correspond to one amplification period in a real system. Pulses will be injected into the loop where, as solitons, they will circulate many times. If those experiments are successful, the day of the elegantly simple, all optical communications system cannot be far off."

HOLOGRAPHY

The concept of viewing a two-dimensional "photograph" that yields a third dimension intrigued the inventor of holography, Dennis Gabor. He conceived the idea in 1947, demonstrated it in 1948, and received the Nobel Prize in 1971 for his efforts. The first demonstration was with ordinary light, but the invention of the laser provided another source: coherent light. Other investigators, notably Emmett Leith and Juris Upatnieks, took an interest in the concept using lasers, and it was not long before this intriguing visual experience of observing the "whole message" (from the Greek *holos* and *gramma* and coined by Gabor as *holography*) was adapted ro research and industrial uses as well as accepted as an art form.

Holograms are becoming so readily available that *National Geographic* magazine has featured them on two recent covers: for the March 1984 issue (a laser-sculpted eagle); and for the November 1985 issue (an image of the skeleton head of Africa's Toung child, claimed to be one to two million years old). A museum in New York City is dedicated to the art of holography. Mass-produced novelties are widely available, even holographic greeting cards. It is possible to computerize three-dimensional holographic signals and plot such images. It appears that 3-D television is not far from reality. By projecting these images, moving objects can be transmitted in space so that future Super Bowls may eventually be played on living room floors. (Demonstrations of this technique were personally viewed at laser exhibits in the late 1970s.)

The following article by the Associated Press appeared in the July 23, 1986, edition of the *Albuquerque Journal.*

> Massachusetts Institute of Technology researchers unveiled a refinement on holography that allows them to project a three-dimensional picture in the air.
>
> Previous holograms were confined to special chambers, usually made of glass.
>
> The MIT scientists said their floating images, much like one depicted in the science-fiction movie "Star Wars," can be designed on a

computer or transferred from sources like a medical CAT-scan to project body organs.

'Using our system, the image is completely projected into space, suspended . . . in front of the observer,' Dr. Stephen A. Benton said of his development, a complex combination of lasers, fiberoptics, mirrors and computers.

Speaking at a news conference, the MIT professor cautioned that the holographic technique was still at an early stage of development and gave no estimate of when the technique would yield practical applications.

He predicted everyday use by surgeons to examine the body and by architects who could have realistic depictions of proposed buildings.

Automobile designers would be able to use computer-generated holograms instead of clay models, he said.

The remaining goal of the three-year, $450,000 research program is perfecting a sharper image in full color and up to 3½-feet long, he said.

At present the holograms are less than 12 inches long and only one hue.

'We have proved that the principle works and we are going on from there,' he said. 'Holographic technology is about where photography was in the 1860s.'

'There have been three-dimensional holograms before,' said Mark Holzbach, an assistant. 'But this is the first undistorted, wide-angle holographic image that's out in the viewer space.

The diagram shown in Figure 6.8 shows light emitted by a laser split into two paths, one to create an object beam and the other to serve as a reference beam. It is the intersecting, or interference, of the two beams and the resultant beam projected onto a photographic film that forms the holographic image. Because every location on the object is illuminated by the object beam, the reflection of these waves intersects the reference pattern and the resulting image appears suspended in space so that as the viewer moves around the image a three-dimensional object is visualized. The two-dimensional representation on the film appears to have depth, with minute detail of recessed surfaces readily observed.

Industrial applications of holography include nondestructive testing (NDT) techniques to find flaws in structural parts. Called holographic interferometry, this method of examining materials for flaws, dimensional exactness, effects of heat or vibration, or other physical aspects has revolutionized NDT, replacing radiography in many instances and supporting other methods of quality control to accuracies not available until the laser was adapted to holography and developed to its present state of the art.

HOLOGRAPHY

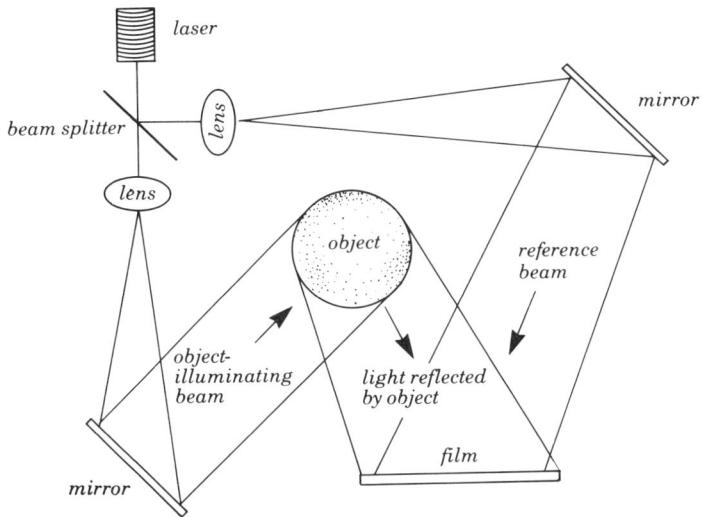

Figure 6.8 Diagrammatic sketch illustrating technique with a laser to produce a hologram on film.

Applications of holographic interferometry have been developed in so many diverse fields that any attempt to include all the current usage would be impossible. However, some of the more advanced techniques with extensive potential in our high-tech society include data storage, whereby written information can be reduced to such small dimension that one source claims that, in theory, all the material in the Library of Congress could be stored on a medium about the size of a regular-size sugar cube. In the operation of a robot, recognition of patterns stored in a memory bank could help in robotic decision making. Holographic logos are being used to foil credit card counterfeiters. Also in this field of identification, holograms are being considered to lend security even to U.S. currency. Sleuthing the paintings of artists by holographic techniques can reveal undercoats.

Although not yet approved by the Federal Aviation Administration, Flight Dynamics, Inc. of Hillsboro, Oregon has developed a device that combines a holographic image of the runway in foul weather and the real world by superimposing a transparent hologram from a mirrored image created by a cathode-ray tube onto the windshield so that the pilot has a "heads up" landing approach.

Holography gives doctors a 3-D view of the inner workings of body organs without side effects. The beating heart, the fetus of a pregnant woman, flowing blood are now available from "motion holography."

An instrument called an ophthalmic laser interferometer has been developed by eye researchers so that the results of certain eye surgery can be predicted. As reported in the March 1986 issue of *Laser Optics*, published by the Laser Institute of America (7), "The pre-surgical prediction of the potential retinal visual acuity allows the clinician to make a more confident qualitative and quantitative determination as to the cost/benefit ratio for the patient." This instrument is also valuable for patients with optical media problems such as lenticular cataract or aberration, or cloudy vitreous which cloud or degrade images. The laser interferometer can also serve as a predictor of multiple sclerosis by determining if the eye being examined maintains the same decimal acuity throughout constant monitoring during a five minute period as a healthy eye would. A suspected acuity would decrease with time.

The ultimate use may be by scientists using holography to help visualize subatomic collisions. It is reported that Fermilab scientists are looking with holograms to find the elusive fundamental particle of matter, the quark.

CHEMISTRY

A laser-based system under development at the Los Alamos National Laboratory has the potential for saving tens of millions of dollars for the hard-pressed US steel industry by speeding up the analysis of molten steel and providing more efficient quality control.

Called LIBS (for Laser-Induced Breakdown Spectroscopy), the system has already been used successfully for rapid determination of the elemental composition of materials ranging from pollutants in coal gasification streams to airborne beryllium in manufacturing environments. A team of researchers from Los Alamos and the American Iron and Steel Institute, with funding from the US Department of Energy, is now trying to extend its capabilities to the analysis of molten material at temperatures in the range of 3000°F.

A commercial Nd:YAG laser—reliable, rugged, and with a good track record in demanding military environments—is used in the LIBS system. Its burst of light, lasting only 10 billionths of a second, creates a tiny fireball of plasma, reducing the material being analyzed to its basic elemental components with characteristic atomic spectra that can be read almost instantaneously on a multichannel analyzer.

Currently the researchers at Los Alamos are devising ways of introducing the laser beam into vats that hold as much as 300 tons of molten steel while avoiding the floating slag at the top of the ladle. If this can be done successfully, LIBS will analyze the steel's composition in less than a minute—at least ten times more quickly than conventional analysis techniques, which involve drawing samples from the ladle and cooling them before transporting them to a laboratory for standard analysis. Because elements are added to steel to form specific alloys for various applications, rapid determination of the ratio of elements in the molten material could speed up production times significantly and improve the quality of the end product.

When the LIBS technique for ladle analysis of trace elements is in place, the researchers intend to go after the big savings by tackling the analysis of molten steel in the furnace. Continuous monitoring there could pinpoint exactly the required heating time of the steel, resulting in enormous savings of both time and energy. Once developed, the spectroscopic method will be capable of analysis for other molten metals and industrial processes.

The function of laser technology in chemistry has increased steadily since the early 1960s when research with gas and chemical lasers was initiated (see Fig. 6.9). Because each laser developed had a specific spectral location, and because chemical spectroscopy had been developed to a mature science, the increased spacings of the electromagnetic spectral lines by this new technology was so dramatic that a revolution occurred in chemical analysis and in isotope separation. The following information from the Los Alamos National Laboratory's *Newsbulletin*, dated January 1985 and written by Jack Challem, describes the progress now taking place in the Laboratory's plutonium chemistry program.

> Los Alamos scientists recently witnessed a major advance in solid-state laser engineering for the Los Alamos Special Isotope Separation (SIS) program. In qualification tests at the end of November, a 100-watt alexandrite laser—a key instrument in the SIS program—exceeded criteria established for its performance.
>
> The principal objectives of the Los Alamos SIS program are, one, to demonstrate the first large-scale separation of plutonium isotopes using the molecular laser isotope separation process and, two, to produce special isotopes in support of the Lab's weapons program. The successful test of the laser ended a multiyear program in which Lab divisions and private contractors progressed steadily toward the completion of an operational high-power laser. The program should lead to a cost effective and efficient means of producing specific plutonium isotopes.
>
> The breakthrough occurred in tests at Allied Corp., a subcontractor in Westlake Village, Calif., from Nov. 27 to 30. Mike Sorem of Tunable Lasers and Applications (CHM-6) and other Laboratory researchers were present during the test. Their plan and that of Allied's researchers called for the laser's sustained operation for six hours at 100 watts with only minor adjustments. The laser operated at specifications with little or no adjustment for 12 hours. Overall, it either met or exceeded established criteria for wavelength, tunability and stability.

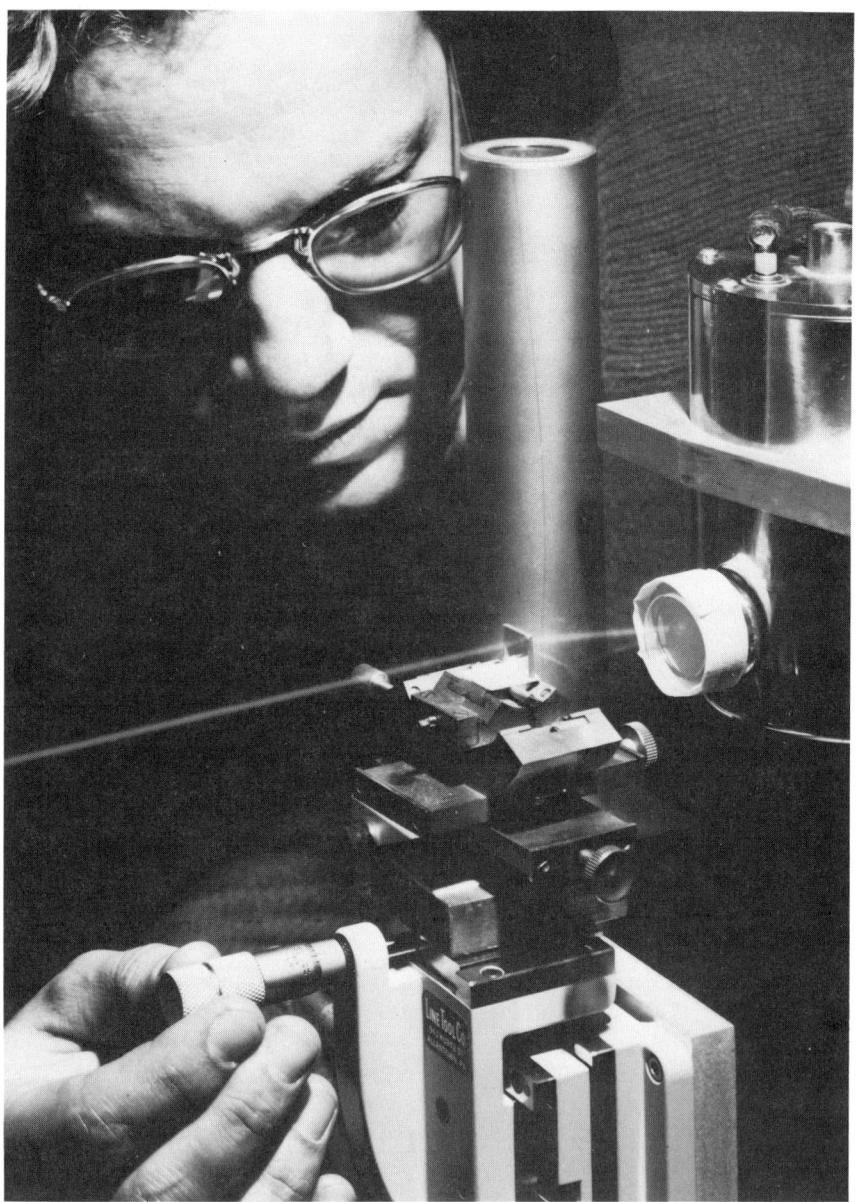

Figure 6.9 Photograph of laser experimentalist adjusting alignment of a HeNe laser beam through a zinc selenide crystal in preparation for experiment in nonlinear optics. (Courtesy of Los Alamos National Laboratory, Los Alamos, New Mexico.)

CHEMISTRY

The test results are particularly significant because during peer reviews in 1982 and 1984, some researchers expressed skepticism of the laser's suitability for plutonium isotope separation, Sorem said. Doubts were raised about the laser's efficiency, thermal-loading limitations, wavelength stability and optical damage caused by the laser beam's inherent high power.

Dick Burick, SIS programs manager, believes that the Los Alamos-Allied prototype is the world's highest average power, narrow bandwidth, tunable solid-state laser.

"Actually there are two objectives in this work. One is to produce the first large-scale demonstration that plutonium isotopes can be separated with the MLIS process. Once we demonstrate that, we will use the facility to produce special isotopes in support of the Los Alamos weapons program," he said.

The laser, arriving in Los Alamos from Allied in February, will become part of a prototype facility at TA-55 to separate specific plutonium isotopes from gaseous plutonium hexafluoride (PuF_6). The process, which is similar to the uranium MLIS process, enlists a variety of exotic equipment including two types of lasers. As it is designed, explained Jay Fries of Chemical Process Development Group (CHM-3), "PuF_6 gas will be pumped by three compressors through a recirculating flow loop in the Laboratory's plutonium facility. The PuF_6 will be mixed with an inert carrier gas, then expanded through a supersonic nozzle that cools the gas."

The alexandrite laser, operating in the near-infrared spectrum, will irradiate the gas and excite the PuF_6 molecules. The alexandrite laser is tuned to excite a specific isotope. Other lasers will then "dissociate the excited PuF_6 to form plutonium pentafluoride (PuF_5), which is a solid. The solid PuF_5 powder is then mechanically filtered from the gas stream," Fries said.

Researchers required a laser with "rather tight specifications on band width, operating wavelength, tunability of wavelength and output power," explained Soren, who is SIS-III laser systems manager. "The medium chosen was alexandrite crystal, which is chromium-doped chrysoberyl. The active ion is chromium, which is why it appears red. Alexandrite is one of the most robust laser materials developed to date. Its strength and thermal conductivity allow it to be pumped to higher average-power levels than any other solid-state materials used to date."

In 1980, the Laboratory began working with Allied, which has a crystals research center, to develop alexandrite lasers for isotope separation. A contract was let in April 1983 to build the prototype that is now being prepared for shipment to Los Alamos.

The laser's tunability from 740 to 800 nanometers is advantageous, Sorem said, because "it can selectively excite specific isotopes—we can tune the laser to any isotope we want to enrich."

Research in other chemistry-related laser technology is progressing at a rapid pace. The effort is so broad and specialized that some conferences are held to disseminate research results in such topics as "Laser Spectroscopy and Clusters," "Laser Photochemistry," and "Applications of Laser Raman Spectroscopy." What some of this specialized research hopes to learn is disclosed in the following progress reports from the Annual Review of the Chemistry Division of the Los Alamos National Laboratory for October 1982-September 1983.

(1). Experiments testing the feasibility of using laser-induced breakdown spectroscopy (LIBS) for real-time detection of airborne beryllium particles were completed this year. Two methods were investigated: (1) direct sampling of air; and (2) sampling of beryllium collected on filters. In both cases the laser spark vaporized beryllium particles and then excited the resulting atoms. Spectral analysis of the spark light reveals the amount of beryllium present in air or on the filter. Results indicate that LIBS would be most useful to analyze beryllium particles collected on filters for two reasons. First, preconcentration of beryllium on filters greatly increases the detection capabilities of LIBS, and secondly, filter collection of airborne beryllium already is used in current monitoring methods. Direct analysis of the filters with LIBS would involve only a change in the analysis method, not the way in which samples are obtained. In addition, LIBS would provide a near real-time analysis of the filters in 5 minutes compared to 8 hours for the currently employed technique, which involves chemical ashing of the filter.

(2). A continuously operating discharge (COD), laser-induced plasma can be generated by focusing the output of a sufficiently powered CW CO_2 laser in various gases. The plasma resides near the focus of the CO_2 beam and has the appearance of a white, very bright ball of light, about 1 to 2 mm in diameter. Because the plasma is maintained by optical radiation instead of frequencies (as in the inductively coupled plasma) it can become very hot. Temperatures of 20,000°K have been reported by others for the plasma sustained in argon gas. Because of the high temperature of the COD, investigating it as an excitation source for emission spectrochemistry was of interest.

(3). The aim of the fluorine atom detection project is to use state-of-the-art infrared detection techniques to determine the minimum detectable concentration of fluorine atoms. The fluorine atom ground state is an inverted ^2P state with a 404-cm^{-1} separation. Nuclear hyperfine selection rules further split this transition into three lines. We used the absorption of a diode laser by the strongest line of the triplet at 404.116 cm^{-1} to measure the concentration of fluorine atoms produced by a microwave discharge in a mixture of F_2 in helium. Using a Cu:Ge detector, we were able to establish that 7×10^{11} F-atoms/cc could be detected with a signal-to-noise of 1. A background-limited detector is currently being tested for signal-to-noise ratio. Preliminary indications are that there will be a significant improvement over other detectors.

(4). The use of ultrafast spectroscopy for the analysis of biomolecular switching devices uses picosecond spectroscopic techniques and five pure functionally interacting photoreceptor proteins to ascertain the role of rapid molecular motions involved in protein-protein signal response mechanisms.

Experiments employing the picosecond laser spectroscopy laboratory include (1) measurements of time-resolved fluorescence polarization anisotropy as a probe of molecular motions and (2) excited-state lifetime measurements to determine intermolecular protein distances using Forster energy transfer theory.

The distance between specific sites on two different interacting proteins can be measured by fluorescence energy transfer techniques. The effect shows biological specificity in that energy transfer is observed only under the same conditions that support biochemical signal transduction.

(5). The production of (boron-free) pure silane by photolysis of impurity materials with the ultraviolet emission from an ArF laser was investigated. The ArF laser selectively dissociates the impurity species arsine, phosphine, and diborane. Construction of a system to purify large quantities of electronic-grade silane was completed.

Integral purification experiments were begun last April, and recently two cylinders of purified silane gas were shipped to Gulf General Atomic for impurity analysis.

In the course of the purification runs there have been problems associated with the formation of a polymer coating of hydrogenated amorphous silicon on the inside surface of the input window of the photolysis cell that reduces the efficiency of the photolysis laser. The formation of this coating may possibly be avoided by substitution of an inert diluent for the hydrogen in the silane process gas. Experiments to study the effect of this modification are under consideration.

These examples of applied laser chemistry are only a few of a plethora of ongoing experiments.

ENGINEERING AND SCIENCE APPLICATIONS

A new publication was announced as this writing was going to press: *Laser Chemistry*. The publishers, Harwood Academic Publishers, NY, in a full-page advertisement in the March 1986 issue of *Lasers and Applications* (2), claimed the multinational journal to be the first journal to

> unite international research activity in this rapidly developing field. It is intended to bridge the gap between physics and chemistry laser-related research. The journal publishes original theoretical and applied papers as well as overviews covering a large number of topics but with particular focus on laser-selective chemistry; relative scattering and state-to-state dynamics; multiphoton dissociation, ionization and absorption phenomena; laser-assisted collisions and reactions in bulks and on surfaces; primary processes in photochemistry and chemiluminescence; molecular relaxation process; dephasing, energy redistribution and ultrafast dynamics; new laser techniques; and applications of lasers in biology.

CONSTRUCTION INDUSTRY

The usefulness of the laser was recognized in the early development of low-power CW units emitting visible wavelengths, especially the red beam provided by low power HeNe lasers. Figure 6.10 shows a battery-powered device in use at a construction site. The uses of these units in all phases of construction details include pipeline installation, whether for exact elevation or for control of sewerline slope for gravity drainage. Building trades use the laser beam to replace the mechanical methods for wall-spacing control and for verticality of block walls and foundation footings. Ceiling levels are controlled by installation of panels to the height indicated by a "sheet" of laser light provided by laser optics. Ground levels and sloping of land for drainage does not require a surveyor's level or transit. Underground construction, especially in the mining industry, has replaced the magnetic compass with the laser beam processed by mirrors linked to surface references. Farmers use lasers to level land and provide exact drainage patterns. Application of the laser in the construction industry appears to be limited only by the builder's lack of ingenuity, or by fear of possible injury to the general public or fellow workers. If procedures are followed as described by the laser manufacturers and a knowledge of laser safety (described in Chapter 5) is acquired, these fears should be allayed.

COMPUTERS AND PRINTERS

IBM has developed an experimental microchip which the company claims will allow computers to exchange data 16 times faster with printers, terminals, and

Figure 6.10 Battery-powered laser at a construction site. (Courtesy of Optomec Design Company, Los Alamos, New Mexico.)

other computers. Using fiberoptic lines (the size of a thread), the chip can process 400 million bits of data per second—more than 17,000 typewritten pages of data.

Lines are of optical fibers and the communications link transferred laser pulses—the fastest rate of signal processing available. Computer-printer industry laser technology makes use of "injection lasers." These are tiny devices, on the order of twelve thousandths of an inch (0.012") used in telecommunications equipment and as "needles" for reading video and compact audio discs. AT&T

claims to have an experimental lightwave multiplexer that combines ten distributed feedback lasers and demonstrates the feasibility of sending 20 gigabits per second through a single optical fiber (see Fig. 6.5).

Sandia National Laboratory in Albuquerque, Nex Mexico, has unveiled the world's first solid-state injection laser made from a strained-layer superlattice (SLS) crystal, a new class of semiconductor material pioneered by Sandia.

But the Sandia laser was not built so much to do a job as to demonstrate the versatility of SLS materials. "This is one of the most difficult semiconductor devices to fabricate," according to Roger Chaffin, manager of the device research department. "If you can make a laser, you can certainly make a transistor."

Semiconductors such as silicon are used to make a wide variety of electronic devices. Strained-layer superlattices are unique in that they combine extremely thin layers of different materials into a single crystal, allowing researchers to tailor-make semiconductors with just the properties they need.

A difficulty arises, however, because the spacing between atoms in each of the layers differs slightly from the layers above and below it. To create the single crystal necessary for proper electrical performance, each layer must be stretched or compressed a bit—"strained," that is—to make the atomic spacing uniform.

When Sandia announced that it had successfully made some SLS materials, skeptics questioned whether a strained material could last long. "We didn't believe them then and we still don't," Chaffin said.

That is where the significance of the SLS injection laser comes in. Though small in size, they must handle a relatively large amount of power—the equivalent of several thousand watts per square centimeter. An injection laser made with a poor crystal simply will not work—or will burn up.

Initial lifetimes have been quite good, Chaffin said.

Some questions remain regarding the durability of SLS materials, acknowledged Ralph Dawson, supervisor of Sandia's compound semiconductor research division. But it's clear they do not self-destruct, and it should be no problem to use the materials for low-stress applications such as light-emitting diodes and light detectors.

"The more we work on this, the more promising it looks," Chaffin explains. "For the world to believe us, we have to stop and show them something (such as the laser) once in a while."

More than a dozen labs, including AT&T Bell Laboratories and the University of Illinois, now are working with SLS materials. Sandia and others have created SLS lasers in the past, but they were laboratory curiosities that were pumped by intense light sources rather than by electricity as is done with the injection laser.

Sandia's injection laser is made of indium gallium arsenide and gallium arsenide and emits infrared light. Different light wavelengths can be attained by using different materials to make the SLS crystal.

Chaffin explains that SLS semiconductors will require at least two or three more years of development before they are ready for commercial manufacture.

With present-day crystal-growing techniques, commercially produced SLS devices probably would cost tens or hundreds of dollars apiece. That's OK for lasers and similar devices, but not for devices such as diodes that typically cost dimes or quarters, at this time.

In the reproduction printing industry, or reprographics, lasers are influencing the market notably. In *Lasers and Applications* (2) January 1986 issue, a review and forecast of the marketplace for lasers included remarks about

> ... the phenomenal success of the diode laser printer for personal computers and office automation. These devices use diode lasers to scan a digitized signal across a photosensitive drum, causing it to lose charge in the scanned areas. Then dry toner adheres to the drum, creating the image when it is thermally bonded to the paper.
>
> Though the print quality is not nearly so good as that from a (laser) typesetter, the price is right and the copy acceptable in growing applications like corporate and newsletter publishing.

In the color printing industry, the report continues,

> Lasers produce the film negatives needed to print color in books, newspapers, and magazines. Using filters and photomultipliers, color separators 'separate' the image to be printed into the four primary colors—yellow, cyan, magenta, and black. Then a modulated laserbeam, usually from an air-cooled argon laser, scans the photographic film to produce the series of four negatives used to print the colored image.
>
> Encouraged by the flashy color graphics in *USA Today*, newspapers around the country have increased their use of color, a trend that should continue for awhile. However, press capacity for color printing, rather than a shortage of color separators, is the limiting factor here, so we expect only moderate growth in sales of lasers for color separators through 1986.
>
> The use of argon lasers is growing in one related category, laser photoplotters. Excellon, Gerber, and Optronics International are selling argon laser photoplotters to the printed circuit board industry. These $100,000-plus machines interface with CAD/CAM computers, greatly improving speed and flexibility in creating the film negatives used to expose photoresist in printed circuit manufacture. Excellon recently has gone a step further: its Direct Imaging System writes directly on the photoresist, eliminating several steps in the process.
>
> Argon and Nd:YAG lasers can be used to burn the plates used on photo-offset printing presses, though this application has always

had more potential than actual success in the marketplace. Almost all of the new systems being sold today use air-cooled argon lasers, though some replacement tubes are sold for older, water-cooled models. Argon lasers expose the plates photographically, whereas YAG platemakers burn the image directly onto a thermally sensitive plate. Crosfield Data Systems is alone in the YAG business, and ion-laser platemakers are made by Dow Jones, Muirhead, Eocom Corp., and others.

The potential here is that many visionaires believe laser platemakers hold the key to automating newspaper production; complete, composed pages of copy and graphics could go directly from the editor's terminal to plate, eliminating many costly, time consuming, and labor intensive steps. Or in the case of national newspapers, composed pages from a central office could be sent by telecom or satellite to remote printing plants, where the plates are made. Incidentally, in 1985, the *Wall Street Journal* completed integration of such a system.

In general, plates have been the limiting factor. They either need high power to expose them, leading to short laser lifetime, or they wear out on press too easily. Last year did see some advances in plate materials, however. Polychrome of Yonkers, NY, has introduced a plate for low-power argon lasers said to combine the sensitivity of film with the strength needed to run on web presses. Also, Anacoil has invented a plate coating compatible with the YAG's infrared radiation, according to Crosfield's David Lightfoot. Despite this, laser platemaking has yet to catch on. Laser sales to this market in 1985 were up 16% to $11.6 million, and we expect sales of $13 million this year, a 10% increase.

MICRO-WIRE STRIPPING

Very thin wires, the approximate size of a human hair, are of great importance in microelectronic engineering; they are used routinely in circuitry with digital or analog elements as well as in miniaturized motors and loudspeakers.

Electronic contacts are presently made by spot welding rather than soldering, since a reliable method to strip the insulating material, which is usually polymeric coating, does not exist. The reason is that damage to the core, which is often coated with a thin gold layer to prevent oxidation, has to be avoided; however, the stripped wire is mechanically much more vulnerable, because the coating increases the tensile strength considerably, thus mechanical methods fail. In addition, thermal stripping is impossible without the risk of burning the metallic core.

Lambda Physik, a producer of excimer lasers, has developed a method to vaporize the coating in a noncontact process without heating the metallic core. The wire is stripped completely around the whole diameter simply by directing the beam onto the wire from different directions simultaneously. This eliminates the need to either turn the wire or the laser beam.

With modern commercial excimer lasers, this method is fast, reproducible, and works on all coatings and all thin wires where other laser methods or thermal methods fail.

ASTRONOMERS' MEASUREMENTS

An article in the *Los Alamos Monitor*, dated April 12, 1985, and crediting Walter Sullivan of the *New York Times*, describes information about the data received from Earth-to-moon-to-Earth laser reflections and is quoted in part here.

> By bouncing light from a newly developed laser off reflectors left by astronauts on the Moon, astronomers have taken new measurements of the constantly changing distances between the Earth and the Moon.
>
> The scientists said they believed the measurements, across perhaps 230,000 miles of space, were accurate to within an inch, ten times more accurate than any made in the past.
>
> The astronomers used a special 80-lens telescope at the Mount Haleakala Observatory of the University of Hawaii to receive the new short-pulse laser beams.
>
> The new measurements, combined with those being made with increasing accuracy from Texas and France, are providing detailed records of day-to-day changes in the rotation of the Earth and the slight wobbles it makes as it spins on its axis. They are also recording lunar motions caused by subtle gravitational effects arising from the influences of relativity.
>
> By analyzing the new data, scientists expect to gain better understanding of the forces deep within the Earth that set off great earthquakes, as well as the variations in the Earth's rotation that have been linked to such devastating effects as El Nino, a cyclical warm-water ocean current that is believed to cause weather anomalies. An improved understanding of such links might open the way to predictions. . . .
>
> The laser beam is aimed by a moving mirror that keeps it pointed precisely at the lunar target despite the constantly changing relative position of the observatory and the Moon. The beam's green light is in a part of the spectrum that allows it to pass virtually unimpeded

through the air. Nevertheless, the billion-watt pulses are so intense that they leave a trace jabbing through the atmosphere toward the Moon. Under ideal conditions, the naked eye can see the returning signal.

So narrow are the outgoing and returning laser beams that if one wanders more than a few hundred yards from the transmitting and receiving site the signal can no longer be seen.

The pulses themselves carry enough energy to damage the eyes of aircraft crews or passengers. The Federal Aviation Administration thus requires that the beam not be turned on until the moon is more than twenty degrees above the horizon.

DOPPLER EFFECT MEASUREMENT

When an approaching car blows its horn, the horn's pitch increases as the car comes near and decreases as it passes. That change in pitch, or frequency, is called the Doppler effect. It is an auditory illusion created by changes in the relative position between the horn and the listener.

That change in frequency was scientifically noted in 1842 by the Austrian physicist Christian Doppler, who based his observation on the whistles of approaching and passing trains.

The Doppler effect also occurs with shifts of the spectrum, and the phenomenon has become an important measurement in science, useful for such measurements as determining the movement of stars. A recent experiment at the Los Alamos Meson Physics Facility measured the shift more precisely than had ever been done.

The experiment was one of a series using a negatively charged hydrogen atom beam and Nd:YAG laser. The experiment's real intent was to increase the laser beam's energy to a range normally beyond its limits and then use it in a variety of atomic physics experiments. The measurement of the Doppler shift was one of these experiments.

The hydrogen ion beam, composed of atoms containing one proton and two electrons, was generated at LAMPF and traveled toward a target at eight-tenths the speed of light. The first laser beam intersected the hydrogen beam, tearing off one of its electrons and neutralizing the beam's negative charge.

From the hydrogen atom's point of view, the laser beam's frequency was shifted to a higher frequency: the Doppler effect.

If you imagine the laser beam as a wavy line, think of the hydrogen beam's motion as compressing the laser beam's waves. That compression, in effect, raised the second laser beam's energy from almost five electron volts up to a maximum of 16 electron volts.

"It's really a trick of sorts," investigator Duncan MacArthur said, "because no commercial tunable laser operates at this energy. By using the Doppler shift, we were able to increase the laser beam's energy, and by changing the angle of the laser beam with respect to the hydrogen ion beam, we were able to tune it."

MacArthur measured various features in the high-speed neutral hydrogen atoms and compared them with well-known features of the atom when it was stationary. "A difference would have told us that the relativistic Doppler shift formula was not correct. But we didn't observe a difference. Everything agreed precisely," he said.

The idea for the experiment was conceived at the University of New Mexico, where some of the components were built.

LASER TUNNELLING

The following article appeared in a recent *Albuquerque Journal* edition.

A new technique for installing sewer lines could save the city on orange barrels and save city motorists from the frustrations of the lane closures normally associated with sewer installations.

New technology—lasers—has allowed tunnelling technique to be applied to the city's pipe-laying projects.

Brian Speicher, a Water Resources Department engineer, told reporters at a mayor's news conference that the new method of installing the sewer lines won't cost the city any additional money above the cost of a traditional installation. He said Loomis Construction Co. of Albuquerque, the contractor for a sewer installation project under way on Yale, SE, is trying the method on an experimental basis. It is used extensively in Europe.

LASER ALIGNMENT SYSTEM

The *Los Alamos Monitor* reported the following information in its February 24, 1985, issue.

A superfine alignment system that will accurately position ninety-six separate laser pulses onto a target the size of a grain of salt has been developed as part of Los Alamos National Laboratory's Aurora laser fusion project.

The technique, which has the unique feature of being able to see through electronic "noise" that usually interferes with resolution, could also be useful in analyzing satellite photos, or scanning battlefields through smoky skies, said a Lab press release.

The Aurora project is aimed at using the world's largest ultraviolet laser to blast a tiny fuel-filled target, causing the atoms inside to fuse and release large amounts of energy.

Commercial laser fusion power plants are many decades in the future, but the Aurora system appears now to be a promising alternative for research.

One of the problems has been directing many laser pulses onto the target so that they all arrive at the same time.

The new alignment system, developed by LANL's Bert Kortegaard, uses television cameras to image the pulses and then relies on a computer to analyze them and eliminate electronic 'noise' which would otherwise interfere with the process.

"The state of the art has been aligning twenty-four laser beams in several hours. With this system, we can do ninety-six beams in several minutes, and with greater precision," Kortegaard said.

The ability to pull a signal out from surrounding noise could apply in other situations—such as the battlefield or satellite photo analysis.

LASER DETECTOR FOR SEARCH AND RESCUE

The *Los Alamos Monitor* ran this item in the March 14, 1985, edition.

From an eastern coast city, a small fleet of helicopters is dispatched into the night on a desperate mission: Far out on the chilly Atlantic a freighter has gone down. The survivors are adrift in small life rafts—or worse yet, scattered through the water in their life vests.

If they're not rescued soon from the cold, choppy seas, many of the men won't live through the night.

With today's search and rescue techniques, their chances are slim—especially in the dark, when it's virtually impossible to spot a bobbing raft, let alone a man in a life vest.

But a technician at Los Alamos National Laboratory has proposed a new search and rescue method that would work best at night—using lasers like specially-designed spotlights to scan a large area.

It could also allow search pilots to fly higher and faster than they can now, so they could rapidly cover more ground. And since the proposed system could be designed to sound an alarm when the laser beam detects something—or someone—it would make the pilot's job a lot easier, said William Cabral.

GUIDANCE BY LASER GYRO

Cabral is an electronics technician at the Lab, but he also has an extensive background in laser work. He presented his proposal recently at the Southwest Conference on Optics in Albuquerque.

It's based on a technique called 'lidar'—an acronym for light detection and ranging. Lidar is very similar to radar (radio detection and ranging), Cabral explained.

In radar, radio waves are sent out, contact something solid, and bounce back. The return signal is detected and measured. In lidar, a beam of light is used instead.

The most commonly used light source is a laser—which produces a beam much more intense and concentrated than ordinary light, Cabral said.

In search and rescue, the lidar laser would be directed out of an airplane or helicopter and pulsed off and on very quickly. If the beam hit either a reflector or a special dye, light would shine back up to the aircraft, where it could be detected on a video screen, hooked to an alarm.

SHOCK WAVE DIAGNOSTICS

Scientists at Los Alamos have used a pulsed laser and a gas gun to probe, for the first time, molecules immediately behind a shock wave. The feat is viewed as a landmark in optical diagnostics of shock waves and detonations and has implications for future development of safer and more efficient explosives.

Tom Rivera, project leader, and Ron Rabie, Explosives Technology, describe the program advance as due to modern laser technology. They said the complex series of events that precedes an explosion is very hard to study because of the extreme temperatures and pressures and the speed of the sequence.

It is very difficult to probe events that happen in a billionth of a second at a speed approaching escape velocity from the earth (25,000 miles per hour), they said.

To determine what actually happens in the dynamic environment of a shock wave that initiates the chemical reaction of molecules before an explosive detonates, researchers used a pulsed laser.

GUIDANCE BY LASER GYRO

The following quotation is from the January 4, 1984, *Los Alamos Monitor*, credited to William Broad of the New York Times News Service.

Today the cutting edge of guidance development is occupied by the laser gyro. Rather than relying on the forces of inertia, it measures changes in counter-rotating beams of laser light that flash around in a tight circle. If the laser gyro itself turns a bit, one beam of light will travel slightly farther around the ring in a given instant of time, the other slightly less far. Differences in the time it takes the laser beams to travel around the ring add up to a precise measurement of the gyro's motion.

The advantages of laser gyros are numerous, according to scientists at Honeywell, Inc., which makes ring laser gyros currently used on the Boeing 737, 757, and 767 jetliners. A conventional mechanical gyro works in dramatically different ways at different temperatures and takes some time to reach a stable speed. Laser beams, on the other hand, always travel at the speed of light.

About the size of a hard-cover book, a ring laser gyro also does away with the complicated system of mechanical gimbals that suspend conventional gyros. Laser gyros can be strapped down to any handy surface in a plane, missile, or spacecraft.

"They're cheaper, smaller, and weigh less than the old gimbal systems," said John Gautraud, a vice president of the Northrop Corp., which makes guidance systems for the military. Without moving parts, they are simply more reliable.

At Litton Industries, one of the world's largest producers of inertial navigation systems, Joseph F. Caligiuri, a vice president, said the laser gyro might eventually be "transcended by newly emerging technologies, such as fiber optics, an even more advanced application of light energy to inertial navigation."

The current king of accuracy, however, is not the laser gyro or a fiber optic device but an esoteric creation for the military known as the electrically suspended gyro. At its heart is a hollow beryllium sphere, which has reference marks on its surface and is suspended in a magnetic cradle. Nothing touches it. Even air is removed from the housing in order to reduce friction. As the sphere spins, a beam of light is bounced off its reference marks and thus measures changes of orientation.

This type of incredibly precise gyro is the navigational brains behind the new generation of American missiles—the land-based MX and sea-based Trident.

LASER WIND-SENSOR

A brief note in the September 1983 *Photonics Spectra* magazine reveals this interesting application.

A laser wind-sensor instrument developed by the National Oceanic and Atmospheric Administration (NOAA) could produce huge fuel savings for commercial airlines by enabling them to alter flight routes to avoid headwinds and utilize tailwinds. If proven in an NOAA-RCA Astro-Electronics feasibility study, the instrument, called "Windsat," would be carried on a modified version of NOAA's Tiros-N weather satellite, using a Doppler lidar to measure radiation from a laser beam shot down toward earth and reflected back to the lidar telescope by windborne tracers such as dust in the atmosphere. By 1990, says NOAA, such a system could save U.S. airlines as much as $200 million a year.

FINGERPRINT IDENTIFICATION

This Cox News Service article in a *Los Alamos Monitor* issue in the spring of 1986 sheds some light on crime!

Lasers, harnessed by police scientists to revolutionize the detection of fingerprints, are shedding a new light on crime.

Crime lab investigators now use the coherent light of laser beams to reveal fingerprints hidden on wood, cardboard, cloth, and, in rare instances even skin, where they may be invisible to the naked eye.

In 1983, Georgia investigators borrowed an Army argon laser to find a killer's fingerprints on the plastic garbage bags he tied around the body of his sixteen-year-old victim. In Ontario, where police first used the technique, detectives cracked a drug case by using a laser to detect fingerprints on the sticky underside of black electrician's tape.

In some instances, the laser has illuminated cases unsolved for generations.

On the back of a yellowed German post card, for example, a laser at the FBI's crime lab revealed a forty-two-year-old fingerprint that unmasked a Nazi war criminal and ended nine years of legal squabbles over his identity.

The post card, mailed June 14, 1942, to Nazi SS leader Heinrich Himmler, was from Viorel Trifa, the wartime leader of Romania's "Iron Guard." Trifa had slipped into the United States in 1950, becoming an archbishop of the Romanian Orthodox Church in America. In 1984, he fled to Portugal rather than face prosecution.

"The fingerprint detected by laser was the key," says Dr. Roland E. Menzel, a physicist at Texas Tech University's Center for Forensic Studies in Lubbock, who has pioneered the use of lasers in fingerprint detection.

"Basically, the idea is to induce fluorescence in the fingerprint. You want the fingerprint to glow like a firefly in the dark," he says. "Fingerprint evidence, if you have it, is extremely strong evidence. The reason you don't hear much about it is that very often it results in a guilty plea."

Fingerprints have been recognized as a unique means of identification since the turn of the century. But, until recently, the tools for detecting the whorls, loops, and ridges left by sticky fingers and sweaty palms were limited to a sable brush, fingerprint powder, and a camera.

If prints were not dusted within a few days, they dried out and vanished. On materials like cloth, even the best technician could find no prints at all.

Menzel, however, found that in response to carefully selected wavelengths of laser light, the amino acids and even some vitamin compounds in the grease and sweat deposited in fingerprints often glowed brightly enough to be photographed and enhanced by computer. Human fingerprints are especially sensitive to the blue-green wavelengths emitted by argon lasers.

SOUND RECOVERY FROM ANTIQUE AUDIO CYLINDERS

An article by Mel Reisner of the Associated Press reported this information in the *Albuquerque Journal* of February 12, 1985.

Fingers of light are being used to lift the sound from audio cylinders so old and fragile they could be classified as artifacts.

The two-year-old Belfer Audio Laboratory and Archive at Syracuse University is the first building in the world designed to preserve and restore sound recordings. It is the only place where cylinders—an early form of recording supplanted in this century by phonodiscs, or "records"—are played back using laser beams and fiber optics, says William Storm, director of the lab.

A laser is a high-intensity light beam; fiber optics refers to the transmission of light along a glass fiber, sometimes with impulses generated by a laser.

In either case, their use in sound restoration involves casting a pinpoint of light into the recording groove, capturing the signal and taping it for indefinite storage and replays.

Laser experimentation in other laboratories has been confined to discs, and Storm says no one else in the field was using fiber optics.

SOUND RECOVERY FROM ANTIQUE AUDIO CYLINDERS

"Playing the cylinders with a normal needle literally wears away the record," he says. "It would be like exposing a great painting like the Mona Lisa to harsh light."

Some of the cylinders are almost a century old—their mass-production began in 1888—and their brittle surfaces cannot stand the weight of phonograph needles, which reproduce sound from mechanical energy, or friction.

Early cylinders were wax over cardboard and played for two minutes. The more advanced "blue amberol" series discontinued in the 1920s featured a celluloid material over a plaster-of-paris core and played for four minutes. But they too are now showing their age.

More than a decade ago, Syracuse alumnus Owen Lewis worked on a solution: Scan the grooves with light. Lewis, now a consultant for the federal government, and a Latham, NY, firm known as Mechanical Technologies Inc., have contributed expertise and equipment in experiments with lasers and fiber optics respectively, Storm says.

He says the archive also has two other ways to replay recordings in better condition—the original equipment, if it does not damage the grooves, and magnetic cartridges.

The twenty-two-year-old archive is a branch of the university's library system, with more than 250,000 phonodiscs, 7,000 cylindrical recordings and about 10,000 tapes.

The center maintains a museum for groups that are shown machines such as an 1877 tin foil phonograph, graphophones and phonographs from the 1880s, and a 1906 Victrola. The groups hear tapes of restored cylinders and discs—turn-of-the-century music, speeches, readings, and instructional lectures.

In a 1912 speech off a cylinder, President Theodore Roosevelt thunders: "We need leaders of inspired idealism, leaders who are granted great visions, leaders who dream greatly and strive to make their dreams come true."

Humor and vaudeville sketches make up another category of cylinders.

Most cylindrical recordings were not produced in large quantities, and most in the archive's possession are rare, Storm says. But because there is no worldwide index of sound, it is impossible to determine whether any of the cylinders is one of a kind.

Last year the sound archives of the Library of Congress, the New York Public Library, and Stanford, Yale, and Syracuse universities completed a two-year project to index more than 650,000 78-rpm commercial discs cut between 1894 and the 1950s. Computer-keyed, the records may now be located through a six-category cataloging system.

ENGINEERING AND SCIENCE APPLICATIONS

Storm says the same institutions want to index their 45-rpm and 33-rpm records issued between 1948 and 1972.

LASER DISC RECORDERS AND PLAYERS

Newsweek, in its January 1985 issue, reported this information:

Phonograph technology has endured 108 years of needle and groove. But now the turntable and today's vinyl LP are on their way to obsolescence. Compact audiodisc players, which use a laser beam to read music encoded on a small, silvery disc spinning at high speed, deliver sound with immense dynamic range and a startling purity. With no stylus or magnetic head touching the disc's surface, the listener is spared the faults that plague record players or tape decks—the scratches, the hiss, the 'wow and flutter' of an imprecise turntable.

In the two years since CD players were introduced in the United States, prices have plunged and a new generation of machines is arriving. At the Consumer Electronics Show in Las Vegas last week, manufacturers showed dozens of models of CD players, including units designed for automobiles. . . .

The heart of the CD technology is a tiny semiconductor laser. Music is recorded on a compact disc as a series of billions of microscopic pits in the disc. As the laser beam strikes the shiny aluminum layer, it is reflected back; when it strikes the microscopic pits, it scatters. The machine "reads" this on-off pattern of light pulses, translates it into a digital code—and then into music.

The technology was jointly developed by N.V. Philips, the Dutch electronics giant, and Sony Corp. of Japan; manufacturers around the world have licensed it, agreeing to a common format. As a result, the most obvious differences between CD players are found in the range of programming features they provide. The laser can be cued to play only certain tracks on the disc, skipping a tune or repeating a satisfying solo passage.

At the moment, the most startling innovation is Sony's tiny, portable D-5, a Lilliputian CD player only slightly bigger than the disc itself and weighing just over a pound. Sony managed to cram most of the circuitry onto a single chip. The D-5 is designed to be carried around, though the mechanism is too delicate for joggers. Moving in a different direction, Pioneer has brought audio and video together in a single laser disc player able to play either CD's or laser videodiscs. Because CD's use a digital format, the unit can also be hooked up to a personal computer to feed in text, video, or sound.

Sony and others have also developed rugged, compact CD players for automobiles; a few versions are already on sale. The automotive CD's are a technical challenge: the players must be rugged enough to withstand jarring potholes and climatic extremes. Mitsubishi's auto CD has a small thermal device to control the climate inside the machine and prevent condensation. Soon portable CD's will reach a different market: last week Sanyo showed a prototype of a monster boom-box with AM/FM stereo, cassette, and a portable CD built in.

When the first CD players appeared, some critics complained that the sound was too bright. Ironically, the dramatic improvements brought by CD's also caused problems. With their uncanny realism, the CD's captured unwanted sound that would have been lost in vinyl—the sound of a door closing in the studio or a musician breathing. Now studio engineers are adapting to the new medium and changing their techniques. The manufacturers have refined the circuitry to improve the sound quality in all second- and third-generation CD players with no loss of compatibility. In some machines the filters have been redesigned, and the tracking mechanisms have been improved to perform faster and more precisely. Although audiophiles delight in the arcane technical differences between different CD players, those differences are much smaller than the range of quality in turntables or other audio equipment.

VIDEO DISCS

According to *Lasers and Applications* magazine in its March 1986 issue,

Optical disk technology had its beginning at Philips Research Laboratories in Eindhoven, the Netherlands, in 1969. The field progressed rapidly thanks to the efforts of a small team of pioneering researchers. They were able to show their first success, the video long-play (VLP) disk player, in presentations to the press in 1972. However, commercial products did not reach the market until late 1978, largely due to adverse market conditions, which continued to favor magnetic video recording and playback.

In 1976, Philips started work on the compact audio disk (CD) concept, which was demonstrated to the press in the fall of 1978. Philips then negotiated on standardization, and by 1980, with some valuable contributions from Sony researchers on the electronics, the die was cast for the phenomenal success of the CD format, which reached the market in late 1982.

The enormous storage capacity of optical media has led to adaptations of optical disk techniques for mass data storage. The first of these products are CD-ROMs, which offer about one gigabit on a single side of the 12-centimeter CD-format disk. Work continues worldwide on erasable and rerecordable optical disks, an area in which Philips is heavily involved.

REFERENCES

1. *Laser Focus/Electro-Optics*. Advanced Technology Group, Penn Well Publishing Company, 119 Russell Street, Littleton, MA 01460. (617) 486-9501.
2. *Lasers and Applications*. High Tech Publications, Inc., 23717 Hawthorne Blvd., Ste. 306, Torrance, CA 90505. (213) 378-0261.
3. *Photonics Spectra*. Laurin Publishing Company, Inc., Berkshire Common, P.O. Box 1146, Pittsfield, MA 01202. (413) 499-0514.
4. Ready, J.F. (1971). *Effects of High-Power Laser Radiation*, Academic Press, New York, 1971.
5. Laser Institute of America, *Guide for Material Processing by Lasers*, Paul M. Harrod Company, 9512 Harford Road, Baltimore, MD 21234.
6. *Optics News*. Optical Society of America, 1816 Jefferson Place, N.W., Washington, D.C. 20036. (202) 223-8130.
7. *Laser Optics*, March 1986. Laser Institute of America, 5151 Monroe St., Suite 102 W., Toledo, Ohio 43623.

7
Medical Applications

Frost and Sullivan, a New York market research firm, predicts a 26% annual growth in laser use in the medical field. At this rate, a $260 million market could be expected by 1990. The key advantages to using lasers rather than more traditional methods for surgical treatment include the precise control of the output energy of the devices and the ability to perform internal operations without resorting to major surgery. An example of the latter is the rapidly advancing art of vaporizing blood clots and plaque to unplug arteries. Debris from the reaction can be removed by the blood stream. Another advantage of lasers in medicine is that different wavelengths have different absorptive properties and penetrate human tissue to different depths, as shown in Figure 7.1. Since all lasers cauterize blood vessels, a choice of lasers makes it possible to contribute to a variety of applications. Those wavelengths that are absorbed near the tissue surface, such as ultraviolet (UV) and infrared (IR) laser beams, outside the ocular focus region, are useful in surgery requiring a minimum of tissue removal, such as required in dermatology (removing undesirable skin tissue) and surgery (minimizing evaporation of brain, liver, or other critical organ tissue). The deeper penetrating wavelengths, such as the 1.06 μm generated by the Nd:YAG laser, are used to cauterize large blood vessels such as those in the stomach resulting from ulcer hemorrhages. A pioneer in developing this operation was Dr. Richard Dwyer, who has been performing the procedure routinely in the Los Angeles Harbor Hospital. The laser is assimilated with an endoscope by providing an optical path with fiber optics and using a fast-pulsed beam.

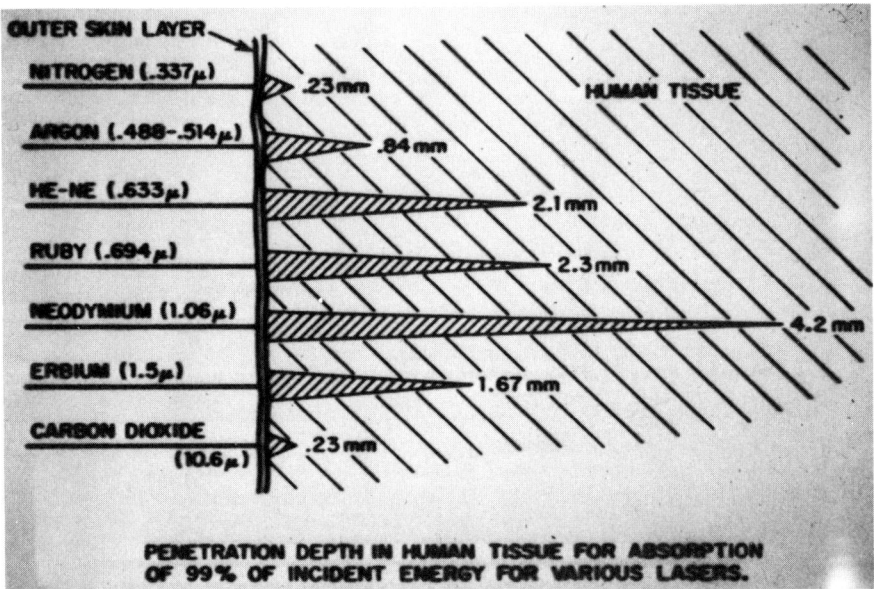

Figure 7.1 Comparative depth of penetration of laser energy in human tissue for various lasers. (Courtesy of R. James Rockwell.)

Another technique involving the use of lasers is the treatment of paralysis, where an HeNe laser is used to stimulate the nerves in the wrists and ankles. Apparently this procedure has the unique ability to stimulate the part of brain that controls motor responses and cause dramatic changes in nerve reactions. According to research conducted on paralyzed human victims by Dr. Judith Walker of Los Angeles and reported to the American Paralysis Association, success has been achieved in restoring hand movement after spinal cord injury, reduction of stiffness and spasms in paralyzed limbs, and a major recovery of body movement in some patients previously classified as permanently paralyzed. Relief of pain by this technique has also been reported.

In this chapter, information has been extracted from recent news articles and other sources to illustrate the adaptability of the laser to various procedures being employed in this fertile field. The applications are so extensive and imaginative that only selected practical uses will be included. Considerable space has been devoted to medical applications because much very innovative work has been done in this field and some of the techniques developed may be applicable to product manufacturing. It would also be helpful for medical laser manufacturers as well as users in considering medical use hardware.

SURGERY

An interesting article on laser surgery appeared in the *Chicago Tribune* dated September 15, 1985. The article, written by science writer Jon Van, follows:

> In medicine as in science fiction, lasers are mostly regarded for their ability to vaporize matter with a concentrated blast of energy.
>
> Lasers reach out to zap enemy rockets in the pages of fiction and in the minds of some military planners. In reality, surgeons use them to burn up brain tumors and remove tattoos.
>
> After years of learning to harness the destructive powers of coherent light beams to serve medicine, physicians are now looking to more subtle uses for lasers. They are experimenting with lasers to "weld" tissue, hoping to bring a new delicacy to surgical techniques.
>
> Laser welding may one day replace needles and sutures for rejoining severed nerves and blood vessels (see Fig. 7.2). It could reduce complications common in brain surgery, and it holds promise for innovations in plastic surgery.
>
> At Northwestern University's medical center in Chicago, laser research suggests that reconnecting severed nerves with laser power may provide consistently better results than with traditional sutures, said Dr. Matthew Quigley and Dr. Julian Bailes, who have conducted tests with lasers on more than 200 laboratory animals.
>
> When a severed nerve is sutured, the stitches hold nerve cells in close proximity so scar tissue can rejoin them. But during the healing process, there is an abnormal "sprouting" of fibers within the nerve that don't rejoin properly. This leads to formation of a neuroma, or benign tumor.
>
> Using a laser to actually melt proteins in the nerve sheath causes the tissue to form a bond that greatly reduces the likelihood of such unwanted sprouting of nerve fibers, said Dr. Leonard Cerullo, chief of Northwestern's neurosurgery and a pioneer in medical laser use.
>
> At Northwestern, where carbon dioxide lasers have been used in the tissue welding experiments, results from reconnecting nerves have been the most promising to date, Quigley said.
>
> Reconnecting blood vessels with laser welding is much quicker for the surgeon than using traditional sutures, Quigley said, but it isn't clear yet if the results are improved.
>
> Another use for lasers Cerullo views optimistically will be sealing the outer covering of the brain, called the dura, after surgery has been concluded. By providing a continuous welded seal around the dura, Cerullo expects the laser will prove superior to sutures.

MEDICAL APPLICATIONS

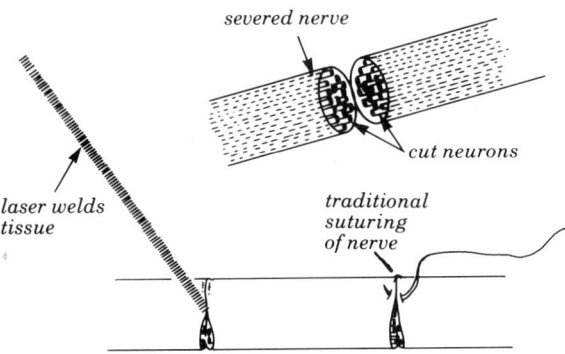

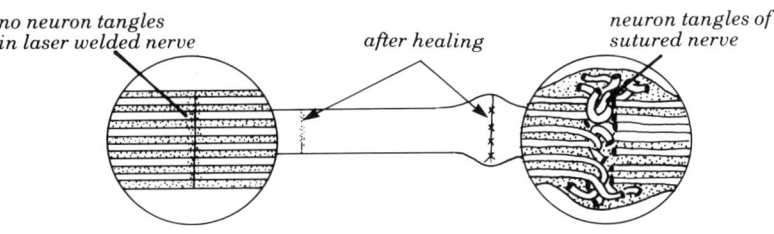

Figure 7.2 Illustrative sketch showing dramatic difference in "welding" nerve tissue with laser energy vs. traditional suturing method.

"Dural leaks after surgery are responsible for an awful lot of hospital days for brain surgery patients every year," Cerullo said. "Eliminating that would be very helpful and cost effective."

The key difference between using a laser to blast away tissue and using it to rejoin or mold it lies in the amount of energy used. A delicate welding job uses about a thousand times less energy than a tissue burning function.

While lasers are becoming common in hospitals, Quigley said, some modifications may be necessary to equip them so they will have an energy range that can both burn and weld tissue.

It is expected that laser welding will result in fewer postoperative infections than sutures because the laser introduces no foreign materials into the body. While Northwestern researchers may be more than a year away from using their new laser welding techniques in human patients, they say this work is one of the most exciting developments in medical laser uses today.

At another Chicago laser center in Ravenswood Hospital, Dr. Rocco Lobraico and colleagues are experimenting with using laser energy to selectively destroy cancer cells while leaving surrounding tissue intact.

About 10 patients have undergone photodynamic therapy at Ravenswood. This involves injecting patients with a purified derivative of a blood pigment that concentrates in cancerous tissue. A laser beam with an affinity for that pigment is later trained on the site and produces a chemical reaction that occurs in the cancer tissue only.

Lobraico said the technique holds promise when used on cancer sites such as those on the face or the vulva that are readily accessible when the cancer hasn't spread to other parts of the body. While such cancers could easily be cut out with a surgical scalpel, the laser technique appears to be quicker, less costly and less painful. It also may provide a better cosmetic result, Lobraico said.

The laser technique apparently doesn't hold promise for advanced cancers that have spread beyond their primary site.

Two advertisements of interest were contributed to this section on laser surgery by Ed Dionne, former Associate Editor of the *National Safety News*, who gleaned them from the September 22 edition of *The Star*, a Chicago area publication. One ad was from a biolaser treatment division of a medical complex offering to remove hemorrhoids and treat other rectal conditions with laser techniques. The other ad, by a group of podiatric physicians, includes a statement that medical scientists have developed laser technology for treating foot problems, and that in most cases, instead of a cut or incision, the the laser beam performs its function by penetration. In describing the laser's versatility, the ad concludes by stating that the procedures have proven successful for removal of ingrown nails, warts, keloids, fungus, nails, planter corns, postop adhesions, and fibrous growths.

The first use of a special laser technique to remove a woman's brain tumor was reported recently in *USA Today*. It shows that surgical methods are catching up with diagnostic techniques, says the doctor who performed the operation. "With CAT scans we can pick up and discover disorders of the brain and nervous system," says Dr. John Tew. "Now we have highly sophisticated devices at our disposal to allow us to treat our patients with tumors and other brain disorders without any lasting deficiency." Mary Boyle, 34, of Cincinnati, was listed in good condition at the University of Cincinnati Hospital following the surgery. Tew received permission from the FDA to use the YAG laser. That agency still considers the laser experimental for neurosurgery. The YAG laser allowed Tew to seal off blood vessels on the tumor's surface while the mass was evaporated with a conventional carbon dioxide laser.

Laser coronary angioplasty, the term for removal of unwanted plaque in coronary arteries, is a technique that is available but needs to be modified for clinical applications. The laser procedure would provide physicians an alternative to open-heart bypass surgery. It would be less expensive and much easier on the patient because opening the chest by major surgery would not be required. According to an article by Jon Van in the May 19, 1985, *Chicago Tribune*,

> There is an urgency in the matter [of developing the laser technique] because the effort to develop coronary lasers also is a race to save thousands of patients who already have received bypass surgery. Those patients have had their blood flow rerouted around original clogged arteries, but in most cases, artery disease continues to narrow the replacement artery.
>
> Once the replacement arteries close down, patients are in serious trouble because repeat bypass surgery isn't feasible, said Dr. Jeffery Isner, a cardiologist at the Tufts-New England Medical Center in Boston.
>
> "We have a large number of patients who have had bypass surgery, and they have disease so diffuse that it won't respond to medication," Isner said. "We have nothing to offer these patients except possibly heart transplantation."
>
> Such patients could be helped by laser angioplasty, which could probably be repeated as needed, once it is developed for clinical use.
>
> Isner's team in Boston and a group at Cedar's-Sinai Medical Center in Los Angeles are at the forefront of experimentation with a laser that offers a wavelength of ultraviolet light that may be just right for the delicate work of cleaning out arteries.
>
> Called an excimer laser, the energy comes from halide gases made from krypton and fluoride or from xenon and chloride. An advantage to the energy beam produced by the excimer appears to be its ability to break apart all materials clogging arteries without producing heat.
>
> The term excimer comes from "excited dimer" and refers to atomic particles formed by a pulsed electrical discharge, which emit ultraviolet radiation as the excited particles relax.
>
> Because the excimer delivers energy in bursts that last a fraction of a millisecond, it offers physicians considerable control in blasting away plaque a little bit at a time.
>
> Working with NASA's Jet Propulsion Laboratory and Bell Laboratory's fiber optics researchers, the Cedars-Sinai team has produced an excimer laser delivery system that contains three fiber bundles [as shown in Figure 7.3]. One bundle shines a light to illuminate the inside of the artery, another contains a tiny lens that sends back pictures to a television screen and the third delivers the excimer energy bursts.

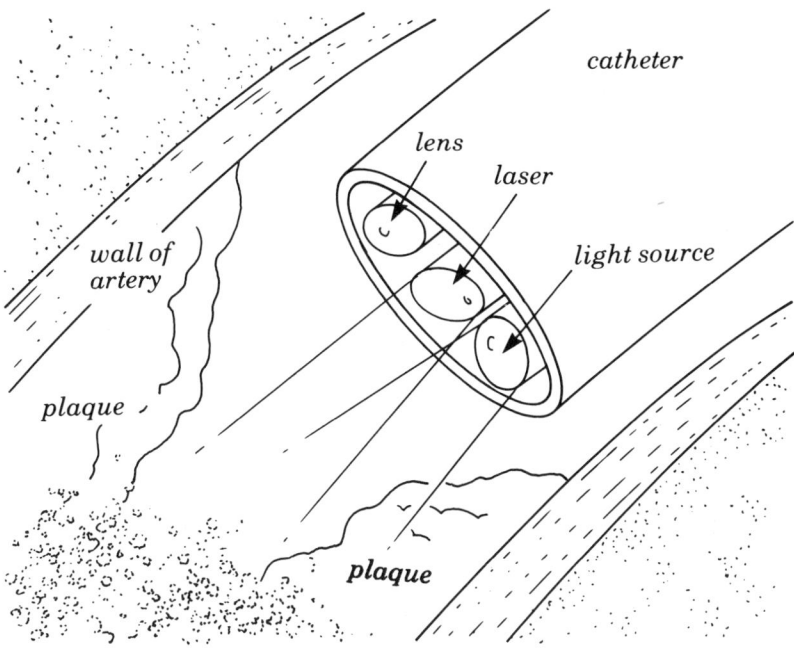

Figure 7.3 Illustrative sketch showing method of vaporizing plaque inside plugged artery using optical fiber to transmit controlled laser energy within catheter.

The whole package can fit into a tiny tube or catheter less than 1.5 millimeters across, but the goal is to reduce that size by one-third. Miniaturization and flexibility are paramount in such a device because of the need to aim the laser at material on the inner artery wall, which moves as the heart beats.

Other researchers, such as Dr. George Abela of the University of Florida in Gainsville, are experimenting with lasers that use argon gas to produce their energy, and they argue that their technology is superior.

But among heart-laser researchers of nearly all schools, there is much agreement that lasers are just short of their promise for heart patients. That last major hurdle must be cleared to deliver on that promise.

What is needed, most heart laser doctors say, is for laser experts from nonmedical fields to devote themselves to the heart laser problem, to transfer technology from space travel, communications and military weapons to this medical application.

"We know the equations work, but new engineering and chemistry is needed to build this from the ground up," said Dr. Warren Grundfest, a surgeon on the Cedars-Sinai heart laser team.

In the experimental excimer his group is using, Grundfest said, each lightwave-carrying fiber costs $10,000, an unacceptably high amount. Yet with proper engineering and wide demand, they should become cheap.

"We're not asking for new technology, just for a medical application of technology already available to military and industrial uses," said Dr. Frank Litvack.

Using their minature fiber optics bundle, Litvack and Grundfest have made videotapes of a journey inside a coronary artery, but such feats may not be enough for precise use of the excimer laser.

The researchers want some kind of feedback sensor device that would detect plaque and monitor it as laser energy is applied to vaporize it. This information, fed to a computer, would allow automatic control of the process so a surgeon wouldn't have to rely solely on his view of a television screen to avoid burning a hole in a patient's artery wall.

The first experiments with lasers using live human patients will be done during open-heart surgery so that it is easier to insert the lasers and to repair any damage they cause. But the goal is to have a tool that is no more invasive than heart catheterization.

Because laser beams cauterize blood vessels, research has been conducted on veins of different sizes, using the appropriate wavelength according to the penetration depth in human tissue. For example, wavelengths in the ultraviolet and infrared regions outside the ocular focus region are the least penetrating wavelengths, so that smaller blood vessels would be more appropriate candidates. Also, these wavelengths vaporize tissue near the surface of the tissue and are used in the careful and minimum removal of valuable portions of critical organs such as the brain and liver. The so-called "laser knife" used in general surgery is the CO_2 beam that requires a minimum of tissue removal. However, in applications that require deeper penetration, such as cauterizing the larger blood vessels found in the stomach lining, the Nd:YAG laser is used in endoscopy. This procedure, developed by Dr. Richard Dwyer of the Harbor Hospital in Los Angeles, as mentioned previously, permits healing of hemorrhaging ulcers that can be cauterized without major surgery by using fiber optical paths to transmit the laser beam within the flexible tubing of the endoscope.

According to a recent Associated Press article from Boston, Massachusetts, laser beams can demolish stones stuck in the body's urinary tract and, coupled with another new high-technology therapy, should nearly eliminate the need for surgery to remove kidney stones, a research reported.

According to the article, the laser technique is the second significant advance in the treatment of kidney stones in recent years. A machine called the lithotripter, which uses shock waves to break up stones, was approved by the U.S. Food and Drug Administration in 1984.

The lithotripter could potentially be used to treat about half of the 200,000 Americans who otherwise would need surgery to remove trapped stones. Now, researchers say, lasers can do the job for virtually all the rest.

Dr. Stephen P. Dretler outlined the experimental use of stone-smashing lasers during a news conference at the annual meeting of the American Urological Association in New York. Dretler, a urologist at Massachusetts General Hospital, developed the technique with Dr. John A. Parrish and colleagues at the hospital's Wellman Research Laboratories.

So far, Dretler has tested the laser on 34 patients, and it has successfully destroyed the painful stones in all but one.

The laser beam, about one-third the size of the lead in a pencil, is flashed through a fiber tube threaded into the body's ureter. The laser emits five pulses a second which chip away at the stone, breaking it up without harming surrounding tissue. Then a tiny basket is used to scoop up stone fragments and carry them out of the body.

Dretler said that, coupled with the lithotripter, the laser should eliminate kidney stone surgery for all but the one or two percent of patients whose stones are too big or hard to be smashed.

About 80 hospitals in the United States already have lithotripters, which pulverize stones lodged in the kidney or upper part of the urinary tract. They cannot be used on stones stuck in the lower part of the urinary tubes. There they are hidden from the lithotripter's shock waves by the bones of the pelvis.

However, Dretler said nearly all of these lower stones can be broken up by lasers. Until now, they would have been removed by surgery.

Neither lithotripters nor lasers are used on stones that can be passed—often painfully—through the urinary tract without getting stuck.

Lithotripters cost about $2 million, and this limits their availability mostly to large medical centers. However, Dretler said equipment for the laser technique will cost only about one-tenth of that and should be available eventually in medium-size hospitals.

The lasers will be tested soon at ten other medical centers in the United States, and Dretler expects them to be approved by the FDA for routine use early next year.

The article concluded that people who undergo kidney stone surgery must be hospitalized for two weeks, and they require 6 to 8 weeks of recuperation.

OPHTHALMOLOGY

In an article in the December 1985 issue of *Prevention**, Kerry Pechter offers "A Consumer's Guide to the New Eye Surgery." The author describes the four most common sight-destroying diseases among people over the age of 45: macular degeneration, glaucoma, diabetic retinopathy, and cataract, and states that new therapies have simplified much eye surgery so that many operations can be performed on an outpatient basis. Much of the following information has been extracted from that article.

Diabetic retinopathy or macular degeneration cause, for complex reasons, the weedlike growth of abnormal, unnecessary blood vessels on or below the surface of the retina—the inside lining of the eye that receives images and sends them to the brain. When these blood vessels multiply or, worse, if they rupture and leak, they can cause blindness.

That is where the laser comes in. It "spot welds" the abnormal vessels before they can do much harm. When the ophthalmologist (sitting opposite the patient in a darkened room and looking into the eyes with specialized contact lenses and a modified microscope) fires either a few or several hundred split-second bursts of blue-green argon laser light into the eyes, each burst creates a tiny burn. The burn dries up the unwanted blood vessel.

This sounds dangerous, but ophthalmologists say it is not. The laser will neither destroy the eye nor burn a hole through the back of your head. To understand why not, think of the way you burned holes in dry leaves with a pocket magnifying glass when you were a kid. The laser works much the same way. It heats only the spot it is focused on, and only for a split second. Think of it as a very sharp cauterizing iron made of intense aquamarine light.

Unfortunately, the argon laser (see Fig. 7.4) cannot help everyone. Some eye patients are better candidates than others. Lee M. Jampol, M.D., chairman of the ophthalmology department at Northwestern University School of Medicine, states that diabetics who seek early treatment have the best chance, but only 10 to 20% of those with macular degeneration can be helped.

"It's important for all diabetics to have regular eye exams," he says. "If we catch the retinopathy in the early stages, we have a good chance of success. The problem is that if you wait until there are symptoms, such as a detached retina or bleeding in the eye, it might be too late." If caught early, the laser can reduce the risk of blindness in diabetics by as much as 70%.

---

*Rodale Press, 33 East Minor St., Emmaus, PA 18049. (215) 967-5171.

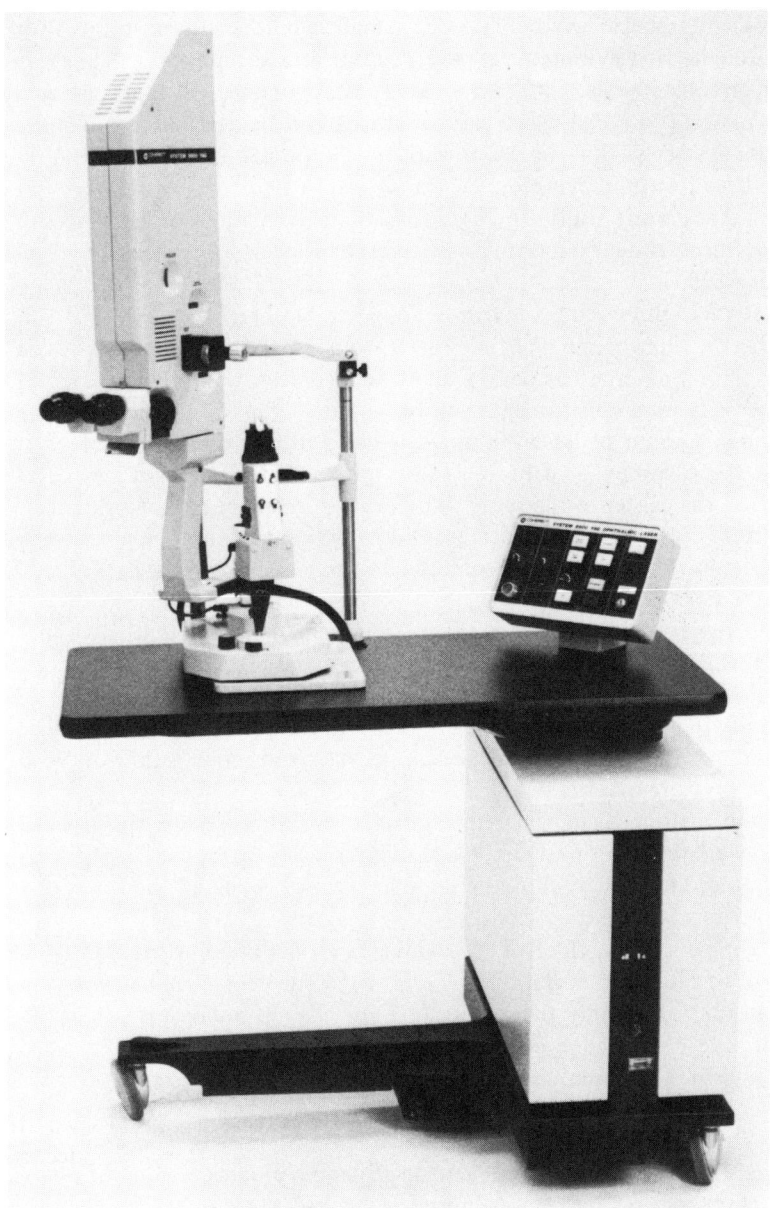

Figure 7.4 Clinical arrangement for use of argon laser in ophthalmology. (Courtesy of Coherent, Inc., Palo Alto, California.)

In macular degeneration, there is only a small "window of opportunity" for treatment after the first symptoms appear. "In this disease a person will develop blurry vision and distortions," Dr. Jampol says. "Straight lines will appear curved. He will lose his ability to read. If we catch it right away, the laser can seal off the abnormal blood vessels. But success is variable. Some people are permanently cured. Some have recurrent symptoms within a year or two."

But, in at least one study, the argon laser cut the amount of vision loss in one group of macular-degeneration patients by half (*Western Journal of Medicine*, February 1984).

The laser does present a few hazards. John H. Mensher, M.D., of the Mason Clinic, Section of Ophthalmology, in Seattle, told us that after extensive laser treatment to the retina, some diabetic patients lose some of their night vision. At times the laser may hit a sensitive spot and cause a stab of sharp, brief, dental-like pain. (If a large amount of laser treatment is performed, a local nerve block is given so that the treatment is pain-free.) In rarer cases, the laser light may damage the macula—the center of the retina—and cause permanent partial vision loss. Also, the various types of laser treatments may temporarily raise the pressure in the eye to glaucomalike levels. Therefore, the doctor may have to check the intraocular pressure after treatment.

Four to five million Americans suffer from glaucoma, a potentially blinding disease in which pressure builds up inside the eye and pinches the optic nerve. The pressure comes from a backup in the flow of fluid. Watery fluid normally enters and exits the eye at all times, but it gets backed up when something clogs up the "drain" called the trabecular meshwork. With too much inflow and too little outflow, pressure mounts.

Three out of four glaucoma patients can control this pressure with special eyedrops. For the rest, an ophthalmologist might suggest 80 to 100 bursts of argon laser. No one knows why, but the laser sometimes opens up the trabecular meshwork and unclogs the drain.

Harry Quigley, M.D., chief of the glaucoma service at Johns Hopkins University School of Medicine in Baltimore, says that some people are better candidates for this kind of therapy than others. "Your chances are best if you are elderly, if you have never had eye surgery before, and if you have what is called primary, open-angle glaucoma," says Dr. Quigley. "In two-thirds to three-fourths of these patients, pressure will go down."

There is a rarer form of glaucoma, called angle-closure glaucoma, for which the laser has also been used. While open-angle glaucoma progresses without symptoms, angle-closure glaucoma causes blurry vision, red eyes, and pain. To treat it, ophthalmologists use the laser to puncture the iris—the colored part of the eye—so that fluid flows freely between two chambers in the front of the eye and pressure drops. "If you catch this early enough, the laser will cure it," Dr. Quigley says.

OPHTHALMOLOGY

Ophthalmologists are now using three new and different surgical techniques to restore vision in cataract patients. The procedures are extracapsular cataract extraction, phacoemulsification, and the YAG laser treatment.

In extracapsular cataract extraction—the most popular form of cataract removal—a surgeon makes a small, curved incision in the eye, reaches inside and removes the front part of the lens, the anterior capsule. This exposes the hard inner part of the cataract, the clouded nucleus.

The nucleus is removed by gentle pressure, causing it to slip out like a watermelon seed squeezed between the thumb and forefinger. A sunction device then removes the rest of the cataract, intentionally leaving the back capsule, or lens-encasing membrane, intact. (This part is used to support the new artificial implanted lens.)

Phacoemulsification is a somewhat controversial and relatively new cataract operation that removes the clouded nucleus a different way. The surgeon makes an even smaller incision in the eye, then inserts a hollow titanium needle into the lens. The needle vibrates 40,000 times a second, simultaneously liquifying the cataract and sucking it out.

Phacoemulsification is attractive because it requires a smaller incision and fewer stitches than extracapsular extraction. But the operation requires specialized training, very expensive equipment, and is more difficult to perform.

The YAG laser (see Fig. 7.5) can be an important addition to extracapsular surgery. In 20 to 30 percent of cataract operations, the back capsule (which was left in to support the new artificial lens) becomes clouded again and obscures good vision. Before the advent of the YAG laser, surgeons were often forced to leave in the fogged-up, vision-reducing capsule or risk a second operation to open it. Now, however, they can use the YAG laser to cut a hole through the capsule in a microsecond.

This "scalpel of light" fires vaporizing bursts that last a billionth of a second or less.

"The YAG causes a mini-explosion at its point of focus," says ophthalmologist Jerome Levy, M.D., of the Manhattan Eye, Ear and Throat Hospital in New York. "Using it is like being able to put a knife in your eye without putting a knife in your eye."

The result is dramatic—a severely visually handicapped person can walk in and leave a few minutes later with near-perfect vision, all because the YAG blasted through the clouded capsule.

The YAG can be used in other kinds of eye surgery, replacing both knife and operating room. Its the most exciting laser now in use. "People say it's a miracle, and I admit, it is a bit of a miracle," says Dr. Levy, whose nurses, no matter how many times they've seen it, still like to go down the hall and watch a patient's transformation.

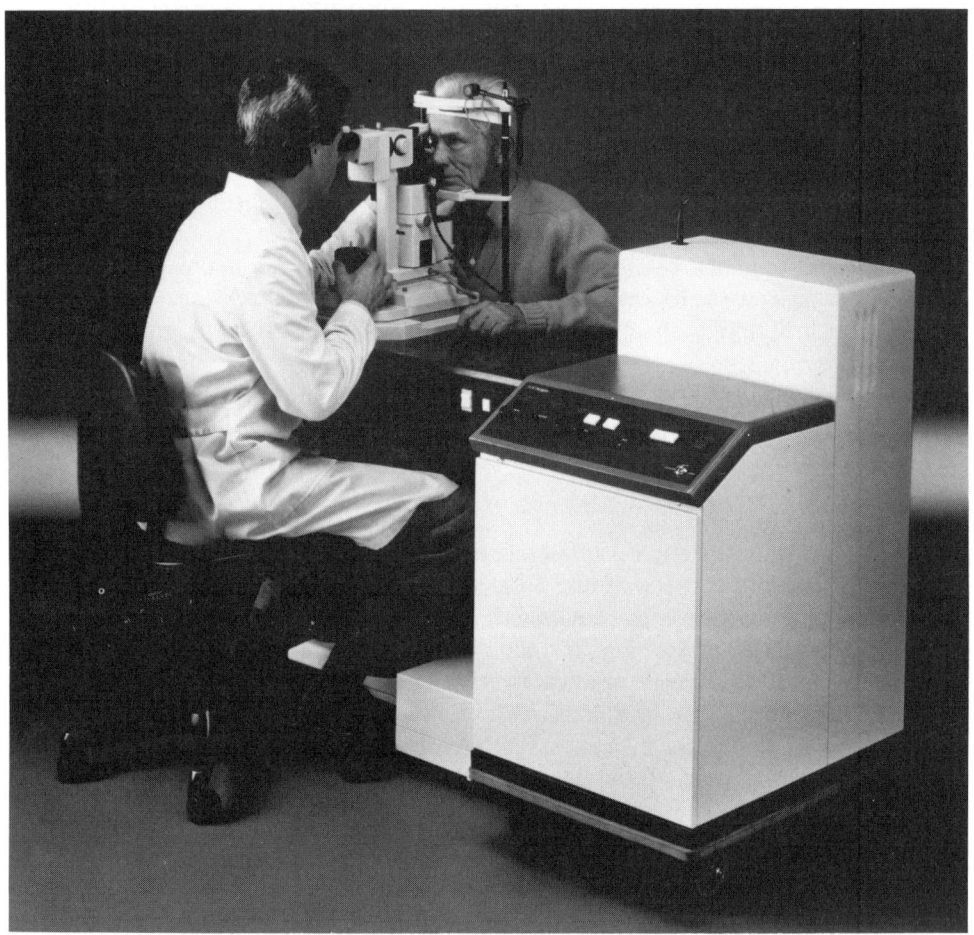

Figure 7.5 Clinical YAG (or argon) laser assembly used in ophthalmology. (Courtesy of Coherent, Inc., Palo Alto, California.)

There are one or two experimental developments in ophthalmology that may make eye surgery easier in the future. One is the silicon lens, now under review by the U.S. Food and Drug Administration. This lens flexes in the middle, making it possible to slip it into the tiny incision the surgeon makes while performing phacoemulsification. One reason phacoemulsification is not yet practical is that the surgeon has to widen the incision anyway for the eye to receive today's inflexible Plexiglas lens implants.

Ophthalmologists have also been interested in something called the excimer laser. According to one doctor, the excimer can cut very precisely and may become the first laser to be used to do radial keratotomy (RK), the newest and most controversial of today's eye surgery techniques. This procedure allows ophthalmologists to treat nearsightedness (myopia) by changing the shape of the eyeball with eight spokelike incisions on the front of the eye. It has enabled some people to throw away their glasses. It has also caused some conservative ophthalmologists to throw up their hands.

Those who have the most to gain from RK, says Beverly Hills ophthalmologist Kenneth Gordon, M.D., are people with mild myopia, who are over 21, who do not have very much astigmatism (an abnormality in the shape of the eyeball), and who cannot tolerate either contact lenses or spectacles.

"People with mild nearsightedness can expect to get nearly 20/20 vision from keratotomy," says Dr. Gordon. "People with more severe nearsightedness—who start out with 20/100 or 20/200, can expect to reach 20/40 vision without glasses. They will probably have to wear glasses for driving or going to baseball games."

The success rate makes this procedure appealing to many.

A study funded last year by the National Eye Institute showed that among 435 people who had undergone this operation, all enjoyed a reduction in myopia. Sixty percent of them had almost perfect vision, and 78 percent had uncorrected vision of 20/40 or better (*Ophthalmology*, February 1985).

The cost of RK is about $1500 per eye. The procedure seems to offer intangible benefits as well. "A lot of people are delighted just to be less dependent on 'a device' in order to live. There's a psychological lift, and people tend to feel better about themselves," according to Dr. Gordon.

There are serious drawbacks, however. It is almost impossible to predict whether the surgery will be successful. Another major consideration is that the incision can weaken the cornea, the lens on the front of the eye. Some people develop astigmatism after RK, and a few become hypersensitive to glare. For some people, vision may fluctuate from day to day. Some experts, including Walter Stark, M.D., chief surgeon at Johns Hopkins Medical School in Baltimore, have publicly opposed RK.

In addition to these recent laser surgical procedures for the eye, reattaching detached retinas with lasers has been an established procedure for many years, as has been the ability to cauterize hemorrhages of the tiny capillary blood vessels in the retina of diabetic patients. As laser technology advances, lasers will find other applications in surgical operations performed on eyes.

DERMATOLOGY

Dr. Leon Goldman, a dermatologist, was a pioneer in developing removal of skin irregularities at the University of Cincinnati Medical School Laser Laboratory.

The treatment of portwine stains and deep-red birthmarks on Caucasian skin has become a clinical procedure. Also, removal of unwanted growths, including carcinomas (skin cancerous cells), have become routine laser treatments.

Laser acupuncture is becoming a popular treatment for some disorders and for the relief of pain. Being painless, fast, and hygienic, laser treatment is becoming more popular in China and Western Europe, the major areas of this practice, than needle techniques.

CELL SORTING

The following information has been extracted from a brochure published by the Los Alamos National Laboratory.

Before the developing of flow systems, researchers examined individual cells and measured cellular components with the microscope—a laborious process. But flow cytometry has changed this process. In flow systems, cells flow rapidly in single file through a chamber where 1000 to 5000 cells are measured each second. Because of their speed, flow systems or rapid cell sorters can analyze many cells and detect the few abnormal cells with high statistical accuracy.

Cell sorting is a multistep process. Either live cells or cells that have been preserved by chemical fixatives can be examined. Sometimes the cells are marked with a fluorescent dye or tag specific for a cellular property. The cells are then suspended in a fluid that carries them through the instrument.

First, the physical and optical properties of the cells are measured. The size and speed of the fluid stream allow only one cell at a time to pass through the sensing region of the flow chamber. Cells leave the inlet tube and are aligned and guided through the first opening, called the "Coulter orifice." Here, the volume of a single cell can be measured using the Coulter principle. An electrical signal proportional to cell volume is generated as the cell passes through the orifice. The signal is translated and stored by a minicomputer for later analysis. After the fluid leaves the Coulter orifice, it enters another sensing region of the flow chamber. As the cell passes through, it is exposed to a beam of light from a laser, and several cell properties can be measured. These measurements can be as simple as measuring the intensity of laser light scattered by the cell or measuring how long the cell is in the laser light.

Flow systems lend themselves to more involved measurements, too. Cells can be stained with fluorescent dyes, which bind to specific parts of cells such as the DNA or protein, or marked with fluorescent tags. Light at the proper wavelength excites the dye molecules or the tag and causes specific parts of the cell to fluoresce—to emit visible light. The intensities of these optical signals are recorded also for further analysis.

CELL SORTING

When the computer receives a light-scatter or fluorescence signal indicating that the cell may be abnormal and might be useful in other studies, the computer initiates a sequence that applies an electrical charge to a series of droplets leaving the flow chamber. These droplets, one of which contains the selected cell, are deflected by charged plates so that they fall into a collection vessel. Those droplets containing cells that do not give the desired signal fall into a separate collection vessel. These sorted cells can then be tested or examined further.

Flow cytometry is not only useful in basic research but is being applied also to specific uses.

Flow systems have been used to detect bacterial diseases (bovine tuberculosis) and viral diseases (Newcastle disease) and to investigate suitable markers for diseases of uncertain origin (cancer). A new method for detecting bovine tuberculosis has developed from flow-system analysis of lymphocytes that have been stimulated by a bacterial antigen from *Mycobacterium bovis*. Cancer has resisted automated detection methods, but studies are being conducted with several fluorescent stains, as well as other markers, compatible with flow systems for their potential use in automated cancer detection.

Researchers at Los Alamos are using high-resolution techniques to detect increased DNA content of lymphocytes—an indication of genetic diseases such as Klinefelter's syndrome.

By analyzing cellular DNA and protein in malignant tumors, we can provide useful kinetic information for characterizing the dynamics of tumor cell growth and proliferation. Researchers are also characterizing two squamous cell carcinomas using automated flow cytometry for growth and kinetics analysis, as well as for enzyme analysis.

Scientists are studying the effects of x-rays, negative pions, and heavy particles on mammalian cells and experimental animals. Los Alamos researchers are also gathering information relevant to clinical radiotherapy about the effects of low- and high-LET radiations on cell growth.

Using flow-system technology, evaluating damage to the immune system caused by exposure to physical and chemical agents associated with the developing oil shale technology is possible. Scientists also are determining the inhalation effects of such toxic agents on the mammalian lung and hematopoietic system. In one study, researchers are evaluating changes in the natural defense mechanisms of the lungs. By measuring the ability of a specific type of cell, called a "macrophage," to ingest foreign particles, we can study the effects of toxic agents on one aspect of the immune response.

One approach of the development of radiochemotherapeutic cancer treatment schedules is to develop a system that can rapidly screen a number of combined radiotherapy and chemotherapy schedules using cells from culture, tumor biopsies, and normal tissues, Also being investigated is drug uptake by cells to determine whether a drug is effective for therapy.

Scientists are developing methods for detecting dicentric chromosomes. New methods are being used to isolate and study abnormal chromosomes that are the result of exposure to radiation and toxic agents.

Researchers are studying the processes of cellular differentiation in populations of normal and cancerous cells. Developing markers to distinguish normal from abnormal cells and developing techniques for discriminating between proliferating and nonproliferating populations is in progress.

Flow systems are beneficial in these and many other applications because they are rapid and extremely accurate, resulting from laser technology.

OTHER MEDICAL APPLICATIONS

Radiologists at the cancer treatment clinic in Santa Fe, New Mexico, use three helium-neon laser beams to ensure the exact location of the area of the patient's anatomy to be irradiated with respect to the beam of the ionizing radiation. The physician marks the desired location with a dye that reflects the red laser beams, and technicians adjust the patient's position until the convergence of the laser beams occurs at the dye spot (see Fig. 7.6).

Woman's World reports in the May 28, 1985, issue that lasers are being used in the treatment of certain arthritic syndromes.

> The same laser beam that prices your groceries in the supermarket may help relieve arthritis pain and promote healing. Robert Willner, M.D., a Miami pain specialist, has been beaming a weakened version of the laser onto the joints of arthritis patients, who report that pain subsides and swelling goes down. How does it work? Injured joints produce prostaglandin E, which causes inflammation and blood clots in arteries around the joint. Dr. Willner theorizes that the laser light converts the prostaglandins to prostacyclin endoperoxide, which unclots blood to promote healing. The laser therapy is now approved by the FDA only for research, not yet for general use.

The development of the free-electron laser by John M.J. Madey and his colleagues at Stanford University from an idea in 1971 to the first working device in 1977 has excited research groups throughout the world who have recognized the potential of the concept, especially with respect to medical applications. In an article by J. Timothy Riordan published in the July 1983 *Photonics Spectra* magazine, the device is called "medicine's rising star." Extractions from that article follow.

> Conservatism is cast to the winds when medical researchers speak of the enormous potential of the free-electron laser in surgery and other therapeutic work. One doctor expects the free-electron laser to be

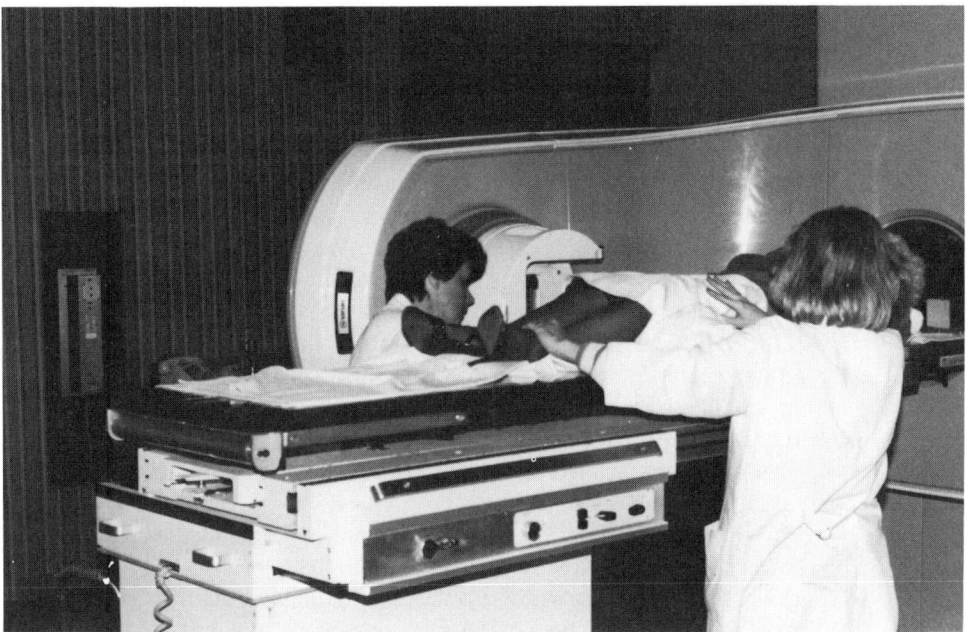

Figure 7.6 Patient in radiation therapy facility being aligned by technicians using HeNe laser beams (3) that converge on dye mark on patient's anatomy. (Courtesy of St. Vincent Hospital Cancer Clinic, Santa Fe, New Mexico.)

"the most exciting medical machine of the next 25 years." Others refer to it as the source of "major breakthroughs" in the treatment of diseases such as cancer, without the liability of harming healthy tissue. All of this flurry centers on a machine that few have seen, let alone used, that still requires considerable development, and that is a rarity even in the research environment.

The free-electron laser is a relatively young device that differs sharply from any other laser type. It requires access to a high-energy electron accelerator, such as the giant Stanford Linear Accelerator. Yet in spite of this cost, enthusiasm has not been diminished. Indeed, medicine is just one of a gaggle of applications cited as promising fields for the free-electron laser; others include spectroscopy, IR imaging, chemical processing, welding and metalworking, laser fusion, communications, and directed energy weapons.

While the high-power applications (especially military) have garnered much of the press for free-electron lasers, Madey has turned his attention toward what he calls the medium-power range—several watts up to several hundred. Not incidentally, this is just the power range that would be appropriate for many of the anticipated medical applications. And Madey says that with current technology, a free-electron laser could be made for $300,000 in parts, or about $1.2 million with labor costs.

In a normal laser, the atoms of a laser medium—crystal, glass, liquid dye, or glass—are excited to a semi-stable state by light or electricity. Light is then emitted by these atoms as they drop back to their ground state, and this light interacts with other atoms in the medium, which release more light. This light of increasing intensity is oscillated back and forth between mirrors until it reaches a threshold level, where it passes through one mirror as a beam of monochromatic coherent radiation.

The free-electron laser uses a very different concept. Electrons from an accelerator are directed into a magnetic field device called a wiggler, which consists of rows of magnets alternating in polarity. As they pass through the wiggler, the electrons are made to vibrate at a specific frequency. Deflection of the electrons from their incoming paths causes them to produce incoherent electromagentic radiation.

To turn these incoherent but monochromatic emissions into a useful coherent beam, the free-electron laser can be operated as either an amplifier or an oscillator. In the first case, an external laser beam, from a CO_2 laser in many cases, is injected into the wiggler along the same path as the electrons. This coherent input acts as a seed for the electron-induced light, causing it to group into a coherent beam. The result is amplification of the CO_2 input. In the second case, the free-electron laser is a self-contained apparatus (if one can use such an expression about a device hooked into a linear accelerator). Mirrors at either end of the wiggler cause oscillation of the electron-induced emissions until they become a coherent output beam of a certain power.

In practical terms, the difference in the free-electron laser's operating principles makes a tremendous difference in its capabilities in comparison with conventional lasers. These differences fall into four main categories:

- broad tunability,
- excellent spot size control,
- excellent pulse width control, and
- high power density and control.

OTHER MEDICAL APPLICATIONS

The tunability of the free-electron laser extends, in theory, from extreme ultraviolet through visible and infrared out to millimeter waves. This tunability can be achieved in two ways: either through changes in the speed of incoming electrons or through adjustments of the distance between magnets in the wiggler.

Medicine has long looked for a means to achieve very specific effects on very specific types of tissue without affecting neighboring tissue. In the most prominent case, doctors would like to be able to eradicate cancerous cells without damaging the rest of the body. What is known about the free-electron laser suggests that it could provide just such a tool.

Kent Sokoloff is a consultant specializing in medical applications of free-electron lasers. According to Sokoloff, the promise of free-electron lasers falls into two primary areas: laser surgery and photoradiation medicine. In the first instance, it is the extraordinary versatility and control possible with the free-electron laser that make it such an exciting development. The ability to produce a very small spot size makes it easier to get at difficult regions and remove diseased tissue with little damage to surrounding cells. In addition, the tunability of these lasers means that they can be adjusted to precisely meet the wavelength needed for a specific effect on a specific tissue. Sokoloff also cites pulse width and power density as critical factors that make the free-electron laser a unique apparatus.

Dr. Dan J. Castro of the Harbor—UCLA Medical Center echoes Sokoloff's enthusiasm about free-electron lasers. Castro is heavily involved in the use of lasers in skin surgery and has specialized in the non-thermal effects of the laser on tissue. He identified four possible results when tissue is exposed to laser radiation:

- the tissue's cells are destroyed,
- the cells will die slowly,
- only some elements of the cell die, and
- there is no destructive effect.

Dr. Castro is quick to point out that these different effects are not strictly related to power output. In fact, his studies show that it is not the coherence and parallelism of the laser beam but its monochromaticity that determines the effect it will have on tissue. Different lesions have entirely different absorption bands, and in fact the same lesion may even have very different absorption regions within itself.

To pin down the effects of a laser on living cells, Dr. Castro has proposed a standard of exposure analogous to the 'RAD,' which is used to measure x-ray dosage. He calls it the LAD (laser absorbed dosage) and says it will be equivalent to the RAD's 100ergs/cubic centimeter level. By incorporating wavelength, power and other factors into a reliable measure of how much laser radiation is actually absorbed by tissue, Dr. Castro hopes to discover scaling laws that will allow prescription of specific laser irradiation for different medical problems.

What does all of this have to do with the free-electron laser? Everything, according to Dr. Castro. This device will let doctors switch from a low power beam at one wavelength that might promote healing to an entirely different wavelength and higher power for surgical cutting. Another plus, he says, is that the wavelength, spot-size and power control provided by the free-electron laser should drastically reduce the number of undesirable side effects of laser surgery.

The second major area where the free-electron laser looks like a winner is in the burgeoning field of photoradiation medicine. In this type of application, photosensitive dyes are injected into the patient and work their way into cancerous tissue. The diseased area is then irradiated by laser output that will trigger a reaction in the dye-saturated cells only. For example, the dye may cause release of free oxygen in the cell, immediately killing it. In short, photoradiation medicine offers a nonsurgical means of knocking out sick tissue, even when it is thoroughly mixed in with healthy tissue.

According to consultant Sokoloff, the free-electron laser is an ideal partner to photoradiation medicine. Again its wide tunability comes into play, but Sokoloff also sees pulse width as a primary factor in photoradiation triggering. In general, tunability in the one- to four-micron range and the ability to generate pulse widths of one to six microseconds are desirable characteristics of a laser in this application, and the free-electron laser meets the requirements. This will allow photoradiation therapists to match laser radiation to the best dye, rather than tailoring dye selection to available laser wavelength, as must be done now.

Another variation of photoradiation involves using dyes attached to antibodies to knock out the diseased portion of a cell. This approach requires an ultraviolet laser and is very dependent upon pulse width, once again calling for just the kind of characteristics a free-electron laser could provide.

John Madey at Stanford produced a report on the development of medium-power free-electron lasers earlier this year. In his report, Madey reviews the design objectives of a medium-power infrared free-

OTHER MEDICAL APPLICATIONS

electron laser, the available and ideal components to be used, and the cost.

Madey sees the priorities of medium-power free-electron lasers dominated by costs, size and complexity, and suggests that because the accelerator-driver will determine these factors, "absolute power output can, and should, be sacrificed if the resulting device can run with a similar or cheaper accelerator." He says that a 10- to 100-watt system with output down to 2 microns could be driven by small conventional pulsed microwave linear accelerators. Using this type of configuration, Madey estimates that a free-electron laser could be designed to be no bigger than current 10-watt argon-ion lasers.

The report cites gain per unit current density as the key design objective for medium-power IR free-electron lasers. In general, it says this type of system—one with optimized small-signal gain—offers the widest tuning range for a given level of accelerator technology and so will keep expenses to a minimum. Another key point: It appears that one micron is the downward limit for free-electron lasers driven by linear accelerators. Thus, without getting into much more expensive technology, the best way to reach into the visible with lower-cost systems is to employ harmonic generation crystals.

Given packaging of a medical free-electron laser in a way that makes it easy for a physician to use and reasonable for installation in the hospital environment, there seems to be little doubt that the free-electron laser will make a major impact on future medical activity. Off-the-record comments by one observer suggest that it would not be surprising to see commercial-scale versions available within two years.

Taking this prospect to the medical laser market in general, this could spell a major upheaval with current product lines. Indeed, one of the reasons the free-electron laser is so appealing is that it could replace the traffic jam of individual lasers now used in medicine for different wavelengths and power outputs. And by extension, a truly versatile tunable laser could send ripples throughout the entire laser market, far beyond medicine.

8
Laser as an Art Form

"Light fantastic" is a phrase used as a humorous reference to an active dancer. It also describes the most exciting visual experiences to illuminate the entertainment scene in recent years: the laser. Laser light shows have been used to complement the music of such diverse groups as the Philadelphia Orchestra and rock groups like Kiss, Wings, and the Electric Light Orchestra. Lasers are becoming routine features of planetariums, discotheques, conferences, amusement parks, state fairs, and even shopping malls.

As beautiful as they can be, high power laser beams can be dangerous if they are not used with a serious concern for safety. Accidental exposure to a high-power laser beam can cause permanent eye damage and severe skin burns. With laser shows that are designed and/or operated by competent and conscientious people, the chance of such accidents is negligible. Unfortunately, however, several light shows have been operated in a haphazard and hazardous manner.

The Food and Drug Administration is the federal agency responsible for protecting the public from radiation hazards from electronic products, including lasers. The FDA's authority includes regulating the manufacture and assembly of lasers, requiring corrective action for those that do not comply with the safety regulations, and educating people about laser safety. Several state radiation agencies are also active in the control of laser products and their use.

Laser products that are in compliance with the established laser standard have certain safety features to reduce the chance of accidents. But such efforts cannot ensure absolute safety. It is up to the laser operator and other responsible parties to see that the laser is used in a safe manner.

Those who sponsor, arrange for, set up, insure, or are otherwise involved with light shows must carry out their part in laser safety.

When a laser beam is reflected off a mirror or other smooth, shiny surfaces, such as water, glass, metal beams, or a glossy floor, it still does not diverge very much. So a reflected laser beam can have almost the same power and potentially the same hazard as a direct laser beam.

Mirror balls are frequently used in light shows to separate and reflect the laser beam into many rays of laser light. When done properly, this can significantly reduce the power and, therefore, the potential hazard of a laser beam. If the beam is reflected off enough facets on the mirror ball, the resulting rays will go off in many directions. Although the individual rays still do not diverge very much, each has only a fraction of the power in the direct beam. Obviously, the degree of risk that this can produce depends upon the power of the direct laser beam, and the number of rays and directions into which the beam is split. The more rays into which the beam is split, the smaller the fraction of power each reflected ray will have. A scanning device is usually used to sweep the beam back and forth across a broad section of the mirror ball so that is it broken up by several facets on the ball. Rotating the mirror ball can provide even more safety because the movement of the reflected rays reduces any exposure time. Without a scanning device, or without a properly designed scanning system, the beam is broken by the mirror ball into fewer rays, each having a larger fraction of the power in the direct beam. This means that even with a mirror ball there could still be a potential for harm.

If a laser beam is reflected off a rough or irregular surface, like a concrete wall or even some "walls" of smoke, the irregularities in the surface scatter the beam in many different directions. The beam is forced to diverge and therefore lose some of its power. However, a very high-powered laser beam can still retain enough of its intensity after reflecting off a rough or irregular surface to cause injury. In addition, some rough surfaces may have shiny spots that allow for a mirror-like reflection of part of the beam.

The question of safety or hazard with laser light shows is "To what levels of power might people by exposed?". The mere presence of a high-powered laser does not necessarily pose a hazard. Scanning safeguards and other measures can be taken to protect people from laser hazards. But, as mentioned above, there is a hazard whenever a high-power laser beam could possibly strike someone, particularly in the eyes. It only takes a fraction of a second to cause serious injury!

This is rationale for government intervention in setting safety requirements for laser light shows. Control of hazards involving possible exposure to the general public are somewhat more stringent than those required in Chapter 5, so additional information on hazard controls is given here.

GOVERNMENT REQUIREMENTS

All laser products made since August 1976 must meet the FDA laser performance standard. Each manufacturer of laser products must report to the FDA about the types of laser products produced.

The standard, based on the ANSI Z136.1 Standard described in Chapter 5, divides laser products into four classes, based on the potential for injuring people and the intensity of the radiation in the laser beam.

Class 1 products produce levels of radiation that have not been found to cause biological damage. Class 1 visible radiation lasers emit less than 0.39 microwatts (or 0.39 millionths of a watt) continuous output.

Class 2 lasers produce radiation that could cause eye damage after direct, long-term exposure. Class 2 lasers emit less than 1 milliwatt (or 1 thousandth of a watt) continuous output.

Class 3 laser products produce radiation powerful enough to injure human tissue with one short exposure to the direct beam or its direct reflections off a shiny surface. Class 3 visible radiation lasers emit less than 500 milliwatts (or one-half watt) continuous output.

Class 4 lasers produce radiation so powerful that it can cause injury with a direct or reflected exposure, even when the beam is scattered or diffused by a rough surface or even by some smoke screens. Class 4 visible radiation lasers emit more than one-half watt continuous output.

All laser products above Class 1, made after August 1976, must carry labels indicating the class to which they belong. Additional safety design and labeling features are required according to the class of the product.

FDA's standard was developed when the use of lasers in the entertainment world was in its infancy. Lasers for demonstration purposes fell primarily into Class 1 or 2 and the standard reflected this. But because of the low visibility of their beams, Class 1 and 2 lasers are not effective with very large crowds. The light shows at concerts and discotheques often use Class 3 and 4 lasers. FDA recognizes that it is possible to use these high-powered lasers in such a way that they will be as safe as Class 1 and 2 demonstration lasers. Therefore, FDA will allow light shows to use Class 3 and 4 lasers as long as the manufacturers can assure safety. FDA does this by means of a "variance." A variance is permission from FDA to deviate from one or more of the requirements of a standard when alternate steps are taken to assure safety. Before May 1980, all of the safety requirements described below were imposed for laser shows except the requirement of an approved variance prior to performance. As of May 1980, the following policy is legally binding: Before Class 3 or Class 4 lasers are sold, used in performances, or otherwise introduced into commerce for demonstration or entertainment purposes, manufacturers must have an approved variance from FDA. Laser manufacturers include people who make laser products and people who receive compensation to design, assemble, or modify a laser projector and/or light show.

This means that a musical group or others are considered manufacturers if they assemble a show ... even if the act of manufacture is simply setting up a show in a particular location or changing a general-purpose laser to light show use, without adding any new laser components. This does not mean that "all the world's a laser manufacturer." And it does not mean that a separate variance is needed for each laser show. But it does mean that, first, the manufacturers of all Class 3 or 4 laser products used in shows that do not already have a variance, must obtain one for each *type* of show performed. Second, all "manufacturers" must submit to FDA a report on all the types of laser products manufactured. A variance must be obtained *before* a laser can be used in a performance or display.

FDA uses several safety criteria to determine whether or not a variance will be granted to a laser light show. These criteria include: The laser must meet all the design and labeling requirements of its class and the following: Laser radiation cannot exceed Class 1 limits where the audience is located. (This can be achieved by proper use of mirror balls, scanning devices, or other safeguards.) If devices, like mirror balls or flat mirrors, are used to reflect the beam, scanning safeguards or other measures are required to make sure that laser radiation above Class 1 will not accidentally go into the audience. Performers cannot be exposed to radiation above Class 1 limits if they must view the laser beam in the course of a performance. When they do not have to view the laser beam, performers cannot be exposed to radiation above Class 2 limits. If the laser is not under the continuous control of an operator, laser radiation above Class 2 limits must be restricted so that it comes no closer than 6 meters (about 20 feet) above, or 2.5 meters (about 8 feet) on the sides or below the floor where the audience would be. If the laser is under the continuous control of an operator, laser radiation above Class 2 limits can come no closer than 3 meters (about 10 feet) above or 2.5 meters (about 8 feet) on the sides or below the floor where the audience would be. Appropriate controls must be taken to make sure that unauthorized persons cannot interfere with the safe operation of the laser. A person must be designated as the laser safety officer who will be responsible for shutting down the laser should any unsafe conditions occur (e.g., should individuals in the audience try to get within the direct laser beam by climbing on a chair or someone's shoulders, or should reflective articles be thrown in the beam). In some situations, as when the audience becomes unruly, strict security measures should be taken to keep the laser operating area free and under the full control of the authorized personnel. Other criteria may be included depending upon the particular show. They may include compliance with state and local requirements, contacting the Federal Aviation Administration for outdoor shows, certification of operators, use of laser cut-off devices or safety shields, time limitations for particular effects, and restrictions on the location of the operator or performers. Once a variance is granted, representatives of FDA must be allowed to inspect the laser equipment and the safety procedures to assure that the conditions of the variance are met. FDA should be notified in writing of all shows at least one month in advance. When this is not possible because of last minute scheduling, FDA should be notified by telephone as soon as possible and then a written confirmation should be sent to FDA.

Anyone who operates laser light shows without an approved FDA variance or who otherwise violates the FDA laser safety standard may be subject to a court injunction and/or civil penalties (fines up to $300,000) as provided for in Section 360C of the Radiation Control for Health and Safety Act. When FDA becomes aware of a particular laser show that is operated in violation of the law or in an otherwise irresponsible fashion, FDA will notify the manufacturer or operator and require corrective action. If the problem is serious, FDA will also notify the state and local authorities and facility managers who can take additional immediate legal steps to halt a hazardous show.

To apply for a variance or for more information about the variance status of a particular laser show manufacturer, reporting requirements, variance applications, and safe operation of laser light shows, contact: Division of Compliance (HFX-430), Bureau of Radiological Health, Food and Drug Administration, Rockville, MD 20857, (301) 443-4874.

OPEN AIR LASER LIGHT SHOWS AND FAA REQUIREMENTS

Even though the chances are small that an aircraft passenger or pilot would be injured by a laser beam from an outdoor light show, the possibility of harm does exist. Therefore, the Federal Aviation Administration must be notified before any open air laser light shows begin operation.

FAA will not object if the output power of the laser beam is less than or equal to one half watt (that is, the laser is Class 1, 2, or 3). As long as aircraft fly no closer than the required 1000 feet over congested areas or over an outdoor assembly of people, there should be little risk from a laser beam of this power. If the show is adjacent to an airport, however, the FAA may object because of the possible risk to aircraft landing and taking off.

FAA will not object to open air shows with Class 4 laser beam powers between one-half and 12 watts if the laser manufacturer/operator informs FAA of the location, time, and laser output sufficiently in advance of the show and if FAA can restrict the air traffic in the area.

In most cases, FAA will object if the laser beam power is greater than 12 watts. A laser of this power is rarely needed for an effective light show and could require extensive restrictions on air traffic.

Notification to the FAA of a proposed open air laser light show should be made in writing at least two weeks and preferably four weeks in advance of the performance. FAA can usually respond with a determination within seven days.

The notification should be directed to the Chief of the Airspace and Procedures Branch at the regional office having jurisdiction over the area where the laser show will take place. The addresses and phone numbers of the appropriate office for each area follow:

Alaskan Regional Office, 701 "C" Street, Anchorage, AK 99513, Tel. (907) 265-4271

Eastern Regional Office, JFK International Airport, Federal Building, Tel. (212) 995-3390

Southern Regional Office, 3400 Whipple Street, East Point, GA 30344, Mail address: P.O. Box 20636, Atlanta, GA 30320, Tel. (404) 763-7616

New England Regional Office, 12 New England Executive Park, Burlington, MA 01803, Tel. (617) 273-7285

Western Regional Office, 15000 Aviation Boulevard, Hawthorne, CA 90260, Mail address: P.O. Box 92007, Worldway Postal Center, Los Angeles, CA 90009, Tel. (213) 536-6186

Northwest Regional Office, FAA Building, Boeing Field, Seattle, WA 98108, Tel. (206) 767-2610

San Juan Area Office, RFD-1, Box 29A, Loiza Street Station, San Juan, PR 00914, Tel. (809) 7911-1250

Rocky Mountain Regional Office, Attn: ARM-530, 10455 East 25th Avenue, Aurora, CO 80010, Tel. (303) 837-3937

Great Lakes Regional Office, 2300 East Devon Avenue, Des Plaines, IL 60018, Tel. (312) 694-4500, ext. 456

Central Regional Office, 601 East 12th Street, Kansas City, MO 64106, Tel. (816) 374-3408

Southwest Regional Office, 4400 Blue Mound Road, Ft. Worth, TX 76101, Mail address: P.O. Box 1689, Ft. Worth, TX 76101, Tel. (817) 624-4911, ext. 306

Pacific-Asia Regional Office, 300 Ala Moano Boulevard, Honolulu, HI 96850, Mail address: P.O. Box 50109, Honolulu, HI 96850, Tel. (808) 546-8354

STATE AND LOCAL LASER LIGHT SHOW REQUIREMENTS

State and municipal governments can have their own requirements, beyond those of FDA and FAA, with which laser light shows must comply when operating in

STATE AND LOCAL LASER LIGHT SHOW REQUIREMENTS 171

their jurisdictions. Presently, six state agencies have specific legislation for lasers and 25 have the authority to develop specific laser regulations. In addition, all states and many local agencies have the authority to take action if a laser show endangers the general health and safety of the public. Several states have closed down laser shows that violated the FDA safety requirements.

The state agencies responsible for radiation control should be notified in advance of laser shows operating within the state boundaries. The following information should be provided in writing by the laser safety officer from either the operating group or the facility where the show is held: name, address, and phone number of laser safety officer(s) or operator(s); name address, and phone number of the auditorium facility and the manager; type of laser show; date(s) and time(s) of performance (if it is not an ongoing show); length of time laser will be in operation; expected attendance; class of laser and name of manufacturer; sketches to describe the design or layout of the show; and, if Class 3 or 4 laser product is to be used, FDA variance and accession number and date of the variance approval.

Since state requirements vary, it is important that laser show operators or facility mangers contact the appropriate office directly to notify the authorities of the operation of a laser show and to ascertain what, if any, additional requirements must be met. Personnel at the state regulatory agency will also be aware of any relevant municipal requirements. Managers of facilities where laser shows are held should be familiar with any local safety requirements. The addresses and phone numbers of the radiation control offices by state follow.

- Division of Radiological Health, State Department of Public Health, State Office Building, Montgomery, AL 36104, (205) 832-5990
- Radiological Health Program, Department of Health & Social Studies, Pouch H-06F, Juneau, AK 99811, (907) 465-3120
- Arizona Atomic Energy Commission, 2929 W. Indian School Road, Pheonix, AZ 85017, (602) 255-4845
- Division of Environmental Health Services, Department of Health, 4815 West Markham Street, Little Rock, AR 72201, (501) 661-2301
- Radiological Health Section, State Department of Health Services, 714 P Street, Sacramento, CA 95814, (916) 322-2073
- Radiation and Hazardous Waste Control, Department of Health, 4210 East 11th Avenue, Denver, CO 80220, (303) 320-8333, ext. 6246
- Radiation Control, Department of Environmental Protection, State Office Building, Hartford, CT 06115, (203) 566-5668
- Office of Radiation Safety, Department of Health & Social Services, Jesse S. Cooper Memorial Bldg., Capitol Square, Dover, DE 19901, (302) 994-2506, ext. 42

Bureau of Occupational & Institutional Hygiene, Department of Environmental Services, 415 12th Street, NW, Suite 314, Washington, DC 20004, (202) 724-4358

Radiological Health Program, Department of Health & Rehabilitation Services, 1317 Winewood Boulevard, Tallahassee, FL 32301, (904) 487-1004

Radiological Health Unit, State Office Building, 47 Trinity Avenue, Atlanta, GA 30334, (404) 894-5795

Bureau of Environmental Health, Department of Public Health & Social Services, P.O. Box 2816, Agana, Guam 96910, 734-9057

Noise & Radiation Branch, Department of Health, P.O. Box 3378, Honolulu, HI 96801, (808) 548-3075

Radiation Control Section, Idaho Department of Health & Welfare, Statehouse, Boise, ID 83720, (208) 334-3335

Division of Radiation Protection, Department of Public Health, 535 West Jefferson Street, Springfield, IL 62761, (217) 782-2342

Radiological Health Section, Indiana State Board of Health, 1330 West Michigan Street, Indianapolis, IN 46206, (317) 633-0150

Radiological Health & Work Disability Section, Iowa Department of Health, Lucas State Office Bldg., Des Moines, IA 50319, (515) 281-4928

Bureau of Radiation Control, Department of Health & Environment, Forbes Field, Building 321, Topeka, KS 66620, (913) 862-9360, ext. 284

Radiation Control Branch, 275 East Main Street, Frankfort, KY 40621, (502) 564-3700

Nuclear Energy Commission, Office of Environmental Affairs, P.O. Box 14690, Baton Rouge, LA 70804, (504) 925-4518

Radiological Health Program, 157 Capitol Street, Augusta, ME 04330, (207) 289-3826

Divison of Radiation Control, 201 West Preston Street, Baltimore, MD 21201, (301) 383-2744

Radiation Control Program, Massachusetts Department of Public Health, 600 Washington St., Rm. 770, Boston, MA 02111, (617) 727-6214

Division of Radiological Health, 3500 Logan Street, P.O. Box 30035, Lansing, MI 48909, (517) 373-1578

Section of Radiation Control, Minnesota Department of Health, 717 Delaware Street, SE, Minneapolis, MN 55440, (612) 296-5323

Division of Radiological Health, State Board of Health, P.O. Box 1700, Jackson, MS 39205, (601) 354-6657

Bureau of Radiological Health, Division of Health, 1511 Christy Lane, P.O. Box 570, Jefferson City, MO 65101, (314) 751-2713, ext. 332

Occupational Health Bureau, Department of Health and Environmental Science, Cogswell Building, Helena, MT 59601, (406) 449-3671

Radiological Health Section, Health Division, 505 East King Street, Carson City, NV 89710, (702) 885-4750

Division of Radiological Health, 301 Centennial Mall, So., P.O. Box 95007, Lincoln, NE 68509, (402) 471-2168

Bureau Environmental Health, Health & Welfare Bldg., Hazen Drive, Concord, NH 03301, (603) 271-4588

Bureau of Radiation Protection, Division of Environmental Quality, 380 Scotch Road, Trenton, NJ 08628, (609) 292-5586

Environmental Improvement Division, Department of Health and Environment, P.O. Box 968, Santa Fe, NM 87503, (505) 827-5271

Bureau of Radiological Health, Empire State Plaza, Tower Building, Albany, NY 12237, (518) 474-2846

Bureau of Radiation Control, NY City Department of Health, 377 Broadway, New York, NY 10013, (212) 566-7750

Radiation Protection Section, Division of Facility Services, P.O. Box 12200, Raleigh, NC 27605, (919) 733-4283

Division of Environmental Engineering, ND Department of Health, 1200 Missouri Avenue, Bismarck, ND 58501, (701) 224-2348

Radiational Health Program, Department of Health, 246 N. High Street, P.O. Box 118, Columbus, OH 43216, (614) 466-1380

Occupational & Radiological Health Service, N.E. 10th & Stonewall Sts., P.O. Box 53551, Oklahoma City, OK 73152, (405) 271-5221

Radiation Control Section, State Health Division, P.O. Box 231, Portland, OR 97207, (503) 229-5797

Bureau of Radiation Protection, Department of Environmental Resources, P.O. Box 2063, Harrisburg, PA 17120, (717) 787-2480

Radiological Health Division, Department of Health, Box 10427, Caparra Heights Station, Rio Piedras, PR 00922, (809) 767-3563

Divison of Occupational Health & Radiation Control, Department of Health, Cannon Building, Davis Street, Providence, RI 02908, (401) 277-2438

Bureau of Radiological Health, SC Department of Health & Environmental Control, 2600 Bull Street, Columbia, SC 29201, (803) 758-5548

Sanitation & Safety Program, State Department of Health, Joe Foss Office Building, Pierre, SD 57501, (605) 773-3918

Division of Radiological Health, Department of Public Health, 344 Cordell Hull Bldg., Nashville, TN 37219, (615) 741-7812

Division of Occupational Health & Radiation Control, Texas Department of Health, 1100 West 49th Street, Austin, TX 78756, (512) 458-7341

Bureau of Radiological & Occupational Health, State Department of Health, Box 2500, Salt Lake City, UT 84103, (801) 533-6734

Division of Occupational & Radiational Health, Department of Health, 10 Baldwin Street, Montpelier, VT 05602, (802) 828-2886

Bureau of Radiational Health, Department of Health, 109 Governor Street, Richmond, VA 23219, (804) 774-6420

Natural Resources Management, Division of Natural Resources, P.O. Box 4340, Charlotte Amalie, St. Thomas, VI 00801, (809) 774-6420

Radiation Control Section, Department of Social & Health Services, MS LD-11, Airdustrial Park, Olympia, WA 98504, (206) 753-3468

Radiological Health Section, Industrial Hygiene Division, 151 11th Avenue, South Charleston, WV 25303, (304) 348-3526

Radiation Protection Section, Division of Health, P.O. Box 309, Madison, WI 53701, (608) 266-1791

Radiological Health Service, Division of Health & Medical Services, Hathaway Bldg., 4th Fl., Cheyenne, WY 82001, (307) 777-7956

INDIVIDUAL RESPONSIBILITIES

Laser manufacturers (including operators or people who set up or assemble laser systems) should have a good understanding of the FDA laser performance standard, the requirements for the class of laser product with which they are concerned, and the safety requirements for laser light show operations.

FDA and state regulatory agency personnel are available to help you make sure that your shows are run safely and in compliance with the law. For more information, contact the appropriate state office or:

Division of Compliance (HFX-430) (see Government Requirements). Bureau of Radiological Health, Food and Drug Administration, Rockville, MD 20857, (301) 443-4874.

Specifically, the following responsibilities are those of the laser manufacturer or operator of light shows:

You must notify FDA, in writing, at least one month in advance of a show. In cases of traveling shows, you may want to send to FDA the schedule for the entire tour. When this is not possible because of last minute scheduling, you should notify FDA by telephone as far in advance as possible and then confirm it in writing.

You must contact the state or local radiation control authorities in writing in advance of conducting a laser show in their jurisdiction. Again, when this is not possible, you should telephone the state agency as far in advance as feasible. In some areas, they also have operating requirements beyond FDA's.

You should also provide to the facility manager the information that is given to the state authority.

If you are responsible for an outdoor laser light show, you must notify the Federal Aviation Administration.

INDIVIDUAL RESPONSIBILITIES

If there are any radiation accidents or alleged accidents, that is, if someone is hurt or an accidental exposure to a laser beam of Class 3 or 4 occurs, you must report the incident to the local authority and FDA regardless of whether any actual injuries occurred.

As of May 1980, if you will be using a Class 3 or 4 laser in light shows or displays, you must submit a variance application and receive an approved variance from FDA before a performance. It is a good idea to have the documents with you during the performance to verify your variance to state or local authorities and facility managers. In fact, this is required in some jurisdictions. Because the public is becoming increasingly aware of potential laser hazards, you may want to include a statement in any promotional advertising of your laser show that it will be operated in conformance with FDA laser safety criteria. If you do so, however, you cannot imply that the show is "endorsed" or "approved" by FDA.

Before introducing a laser into commerce, you must submit to FDA a report describing your laser products and the manner by which they comply with the FDA laser safety standard and the conditions of any variances. This "initial report" must be followed up by a "model change report" should you plan to introduce a new or modified laser show or device into commerce. An annual report must also be submitted by September 1 of each year summarizing the testing and the records that must be maintained.

In order to safeguard audiences, facilities managers should be aware of the safety requirements placed on the manufacturers operators of laser shows by federal, state and local authorities. To avoid possible liability for laser injuries, any shows in the facility must comply with the legal requirements.

NOTE: The laser operator should provide the information about the class of laser to be used and its variance status. A laser product should have a label indicating its class. If the laser is Class 3 or 4, the company responsible for the laser should have documentation (an accession and variance number) from FDA granting a variance. The state authority or FDA should be contacted to verify the status of a company's variance.

One person, either the laser company's operator, or, where there is no operator, an employee from the facility, should be designated as laser safety officer. A laser safety officer should be in attendance whenever a laser is in operation and should be responsible for shutting down the laser should any unsafe conditions occur.

In order to properly set up and align a laser light system that can be operated safely, laser groups will need time in the facility before the show without members of the public present. They will need the electrical power and water supply set up early enough to test and align the equipment. Depending upon the complexity of the system, the preparation for a show may take up to several hours. Should a full inspection by FDA representatives be found necessary, it may require an additional hour or two prior to the show. This should be allowed for in the scheduling of performances.

Should any accident occur with the laser, you should report the incident to the state authority and to FDA.

FDA and state personnel are available to help you ensure that laser shows in your facility are run safely and in compliance with the law. Contact the FDA or the appropriate state office.

WHAT THE PUBLIC SHOULD KNOW ABOUT LASER SAFETY

Laser light shows can be exciting, but they can also be hazardous.

The public has a right to enjoy a laser show knowing that safety is provided for by the laser manufacturer, the laser operator, and the management of the facility where the show is held. Should there be reason to believe that a show is not being run safely—that is, that the precautions described herein are not being taken—observers should talk with the laser operator or people in charge of the facility, or call the state authority. If anyone is aware of anyone being injured at a laser show, it should be reported to the state authority or FDA.

9
Future Developments of Laser Technology

AUTOMOBILE NAVIGATION BY LASERS

A *Los Angeles Times* article by Donald Woutat appeared in the October 21, 1985, issue of the *Albuquerque Journal*'s "Business Outlook" section and described the status of automobile navigation.

> ... People have been trying for decades to come up with a navigational system for a car. One early proposal was a road map in a roller that would scroll as the car's wheels turned. A bell would ring when a certain number of miles have been driven and it was theoretically time to turn.
>
> The auto companies would like to avoid systems that depend on the stars, costly radar systems, easily interrupted radio signals, or other approaches normally used for ships and aircraft. The major companies, including GM, are developing navigators that would rely on radio signals beamed from the expanding number of navigation satellites.
>
> So far, there have not been enough such satellites aloft to offer a twenty-four-hour system. That will change by the decade's end. Chrysler, for instance, hopes to combine signals from navigation satellites with laser disks, which can hold far more map detail than other systems—a single laser disk could cover the entire United States with

177

seven levels of maps, scaled as low as fifty square miles. But the satellite approach assumes the development of a satellite receiver that is small and cheap enough for cars.

Donald Gero, manager of electronic product development at Chrysler's Huntsville, Ala., electronics operations, said the vote is still out on which navigation technologies will prevail in the 1990s.

Robert McMillan, director of engineering at GM's Delco Electronics division in Kokomo, Ind., said GM's Delco Electronics unit, which makes navigation systems for missiles, spacecraft, and ships, came to believe that conventional navigation technology "is not suited for the automobile because it has its wheels on the ground. We really must capitalize on that."

Etak, a Sunnyvale, Calif., start-up company founded in 1983 by three engineers from SRI International, the Palo Alto, Calif., research company, did just that by combining wheel sensors and a compass with a modest computer and the cassette approach to mapping. The result is a dead reckoning type of navigation system that does not rely on any outside signals or require any programming by the driver.

With the help of a substantial infusion of cash from GM and the prospect of much bigger bucks down the road, Etak has brought to market the first self-contained, computer-based device with a small screen that actually shows drivers where they are and where they are going.

They are pushing the navigator as a valuable tool for delivery and rental-car fleets, for sales people, real estate agents, and other frequent drivers, and as the basis for other products.

"All the marketing clinics have been very favorable," said McMillan. "We see the most interest among people who make their living finding their way around."

In about three years, GM expects to install the device in vehicles on the assembly line as the centerpiece of a sophisticated electronic communications system. With GM's sales of more than 7 million cars and trucks worldwide a year, it is easy to imagine a royalty windfall for Etak. GM also has warrants to buy ten percent of Etak. For now, the navigator is making its way into shops in Los Angeles and San Francisco that sell car stereo systems, cellular phones, and two-way radios. As Etak maps other metropolitan areas, a task it expects to finish by the end of next year, it will become available in those communities.

"What they were doing closely paralleled what we felt ought to be done," McMilland said. "We'd been looking at a number of navigation systems when our people heard about Etak. It was a neat way to get a leg up. We fell very confident this is the way we want to go."

MILITARY APPLICATIONS

The following editorial is from the April 1982 *Photonics Spectra* by publisher Teddi C. Laurin. Follow-up articles from various sources will be included to help understand the "progress" being made in laser weapons in space and provide somewhat of a chronological order.

If strategic arms limitation talks between the United States and the Soviet Union are resumed this summer, as now seems likely, it might be a good idea to expand the agenda to include a ban on laser weapons in space. Unless steps toward this end are taken, and soon, there is a very real danger that the present balance of terror will be escalated to a new and more perilous dimension.

We are part way there already. The Soviets, if one can believe the warnings emanating from our Defense Department, are preparing to put a laser weapon in space within the next two years—supposedly the first of a number of laser-armed vehicles capable of knocking out US satellites used for military communications, espionage, and early warning of enemy attack. The assumption is that the Russians will soon be able to orbit a laser powerful enough to do crippling damage to these fragile vehicles, though why lasers would be more logical for this purpose than conventional explosives or pellet barages is not at all clear.

In any event, whether the threat is real or fancied, our own military planners are calling for costly countermeasures. They envision— at an estimated cost of more than $15 billion over the next ten years— the creation of up to a dozen orbiting anti-satellite (A-SAT) stations, also laser-armed, which would have the specific mission of knocking out the Russian A-SATs before the latter could reach their targets. The scheme seems assured of a strong cheering section in Congress, which last year insisted on adding $50 million to the Reagan administration's already generous allocations for space-based laser weapons.

There are other Buck Rogers proposals on the drawing boards. Prodded by defense contractors, a group of congressmen is pushing hard for development of an orbital laser ballistic missile defense, which Republican Senator Malcolm Wallop of Wyoming claims "would revolutionize the strategic equation." It would do so by deploying in space several dozen laser systems supposedly capable of destroying with the speed of light enemy ballistic missiles as fast as they are launched. The estimated price tag is $50 billion.

Whether any of these schemes make sense is debatable. Charles H. Townes, the laser pioneer and Nobel laureate, has called the laser ballistic missile defense "science fiction" at this stage. Recent studies at MIT concluded that "lasers have little or no chance of succeeding as practical, cost-effective defensive weapons." A recent article in *Science* magazine pointed out that military satellites and their laser defenders "could easily be destroyed by a single nuclear blast in outer space."

Sensible or not, however, there is no reason to doubt that both the super-powers will continue their preparations for placing weaponry in orbit unless verifiable agreements to refrain from doing so can be negotiated. The task is urgent and the time is short.

Star Wars Technology

An article appeared in the March 26, 1986, issue of the *Albuquerque Journal* by David Morrissey, headlined, "Los Alamos Scientists Played Role in Birth of Star Wars." The following is quoted from that article.

In 1983, President Reagan stunned the nation and much of America's scientific community with his vision of an anti-missile laser system, called the Strategic Defense Initiative, that would make "nuclear weapons impotent and obsolete."

One of the few places where Reagan's vision wasn't a surprise, and, indeed, was a dream already shared, was Los Alamos National Laboratory.

Los Alamos can claim to be one of the intellectual "fathers" of Star Wars, as SDI is now popularly known. The extent of Los Alamos' involvement in conceiving SDI became clear in a revealing incident the night the president enthusiastically endorsed the new system.

U.S. Sen. Pete Domenici, a key Republican ally of the president, was watching the televised speech. When the president proposed SDI, Domenici excitedly exclaimed to those in the room, "That's our project!"

Personnel from several research facilities, in New Mexico and other states, played a role in convincing the administration SDI was feasible, Domenici said.

But it was Los Alamos scientists, Domenici emphasized, who gave him a report demonstrating how modern optics and laser technology could be used to create a missile defense system. "We gave the president the report on the optic weapons system more than a year ago," Domenici told the *Journal* in March 1983. "Apparently someone has sold him on the system."

From a virtually unknown proposal in 1983, with little funding and few supporters, the Strategic Defense Initiative has grown into an important Los Alamos project. Last year about 12.5 percent of the 1985 Los Alamos operating budget—some $86 million—went for SDI programs, said Los Alamos official Barbara Mulkin.

This rapid increase in funding for a defense project at Los Alamos is noteworthy not only for what this particular program would do but for the trend it represents. Across the board, Los Alamos is becoming increasingly involved in defense research and development— as opposed to research and development for non-defense projects.

Los Alamos officials, while recognizing the trend, say it merely reflects a national realignment of priorities. . . .

In the September 1985 issue of the Los Alamos National Laboratory's *Newsbulletin*, the following information was revealed.

With the achievement of another first, Los Alamos last week took an important technical step in accelerator beam technology. Researchers in Hydrodynamics (M-4) showed the feasibility of a laser beam guiding an electron beam, a concept that may have application in space-based weapons.

M-4 staffers were elated as they guided a powerful electron beam farther than one had ever gone before. Their experiment showed that electron beams from radiofrequency accelerators could have applications in space-based directed-energy weapons and in a future radiography diagnostic facility at Los Alamos.

Randy Carlson and Steve Downey of M-4 combined a simple principle of physics with a sophisticated laser and accelerator to guide the electron beam.

Although the electron beam only travelled eleven feet, the researchers showed that a laser beam could guide it outside the confines of an accelerator and over a distance. In a follow-up experiment, the researchers will attempt to direct the electron beam about forty-five feet, which will be another world record.

Several features distinguished the experiment from similar research at other national laboratories, Carlson explained. First of all, the electron beam was fired in series of pulses rather than in a single burst.

Second, those pulses were at a high energy—27 million electron volts—that was "monoenergetic," meaning tuned to a specific energy range. Such fine tuning would be necessary to focus the beam in any application.

They used the accelerator to produce a pencil-thin radio-frequency electron beam about 1,000 times more powerful than the average medical X-ray. The beam was aligned to enter a one-foot in diameter by eleven-foot stainless steel propagation tube in pulses lasting five-billionths of a second.

But about fifty-billionths of a second before the electron beam entered the tube, Carlson and Downey created a path for it with a slightly wider krypton fluoride (KrF) laser beam.

They knew what just about everyone knows: that like electrical or magnetic charges repel and opposites attract. The laser beam ionized a channel of benzene gas in the tube, creating positively charged benzene ions and negatively charged free electrons. The benzene ions formed a path for the electron beam, drawing it forward.

In the fall of 1985, "the first phase of America's 'Star Wars' defense system was unveiled," according to a recent article in the *Albuquerque Journal* by Joe Smith, "and officials said initial testing has been 'very, very promising.'" Continuing, the article is quoted as follows:

Focal point of the testing has been MIRACL—the Mid-Infrared Advanced Chemical Laser—a device likened by Naval Ordinance Missile Test State Commander Capt. Art Schroeder to an "exotic rocket engine."

But instead of powering missiles into space, this rocket engine is designed to shoot them down.

In tests last month, MIRACL successfully destroyed both solid- and liquid-fueled Titan-1 rocket boosters at the High-Energy laser Test Facility, just west of White Sands National Monument.

The laser, Schroeder said, uses various chemicals to produce a sixty-inch laser beam, which is conveyed by mirrors through a high-precision pointer-tracker called a beam director.

While the system is supported by banks of computers and other electronics, facility director Dr. John Davies said the laser itself is an energy generator, not an energy user, and operates without electricity or any other source of power.

Its power-generating capacity is classified information, but officials did say the device's continuous power output is the highest ever achieved in the U.S.

The tests at White Sands are part of a five-year program ordered by President Reagan in 1983, Davies said. About $300 million of the program's $1.4 billion first-year budget went into the facility, he said.

Schroeder, Davies and Strategic Defense Initiative Organization public affairs officer Lt. Col. Lee DeLorme, Jr., declined to speculate on what White Sands' role will be after that, other than to test various materials.

Nor would they comment further on the laser's other capabilities.

DeLorme said the White Sands tests are designed to prove three things: feasibility, workability, and affordability, especially in the area of "lethality"—the device's ability to knock out incoming enemy missiles and aircraft.

In the January 1986 issue of *Physics Today* magazine, the following article appeared under "Washington Reports."

As 1985 drew to a close, the debate about the President's Strategic Defense Initiative continued to spread and intensify in the physics community. Petitions opposing Star Wars picked up more signatures in physics departments around the country, but some physicists and computer scientists entered the fray on the other side, issuing public statements that strongly supported the SDI research program. Meanwhile, significant changes in the SDI program were announced that seem to indicate that Star Wars will be debated on different terms in the coming years than it was in 1983-85.

In a nutshell, some of the technologies that were much debated in the first phase of the SDI program—high-power space-based chemical and excimer lasers, in particular—are being put on the back burner. Instead, priority is to be given to rapid development of a system of heat-seeking or command-guidance rocket interceptors suitable for deployment in space and on the ground, and to ground-based free-electron lasers.

To experts who had been following the program closely, the changes in the program did not come as a great surprise. In fact, Lieutenant General James A. Abrahamson, the director of the Strategic Defense Initiative organization, had given pretty clear advance notice that a streamlining of the program was to be expected.

In a speech delivered at a space-technology conference in Colorado Springs on 20 November—the day President Reagan was conducting final discussions with Soviet leader Gorbachev in Geneva—Abrahamson said that he expected to receive instructions after the summit to move ahead "much more quickly and effectively" with research into the design of a space-based defense system against nuclear missiles. While Abrahamson's speech was not prominently reported in many newspapers, it was striking because it expressed confidence in the President's intentions at a time when other Administration officials were betraying considerable nervousness about what Reagan might agree to with Gorbachev.

On 26 November, two days before Thanksgiving, it was announced that Abrahamson would hold a press conference following the daily DOD briefing at the Pentagon. This was not billed as an event where significant policy changes would be announced, and to the extent Abrahamson's remarks were reported, they tended to appear on the inside pages of the final preholiday newspapers.

Abrahamson opened the briefing by remarking that he was there "not to talk about some of the policy issues on the program nor to talk about Geneva." What he intended to do, he said, was to give "an overall impression of some of the very steady progress that we are making . . . and some of the neat things that are coming out."

Emphasizing that SDIO remained committed to the principles of a multi-layered defense system and boost-phase interception, Abrahamson proceeded to enumerate in a free and broad-ranging style various developments in the program and problems that are to receive urgent attention. He made, for example, the following points:

The development of effective ground-based terminal and midcourse defenses will depend on better techniques to discriminate between decoys and real warheads.

Some but not thousands of terminal defenses will be needed around certain areas or targets considered particularly important.

There has been "incredible" progress with free-electron lasers at Lawrence Livermore Laboratory, and while such lasers could not be based in space, test results have indicated that they could deliver enough energy from the ground to destroy boosters and that the problem of atmospheric distortion could be overcome.

Test results from the antiballistic-missile "homing overlay experiment" in June 1984 and a recent test of an antisatellite weapon have indicated the effectiveness of heat-seeking interceptor rockets.

A ground-based heat-seeking interceptor rocket is being developed for terminal defense, and under the ABM Treaty "we could go all the way even to deployment for 100 of these if it made sense," though it would be better to wait until there is a basis for determining the viability of a full multilayered system.

Emphasis needs to be placed on development of radiation-immune data collectors and processors containing thousands, tens of thousands, or even millions of elements.

Highly decentralized signal-processing systems are being considered particularly thoroughly.

In addition to the 300-kW reactor being developed at Hanford for use as a space-based power source, other innovative ways of storing and converting very large quantities of energy will be needed.

Cheaper methods of making very large mirrors will be needed if systems based on ground-based lasers are to be viable.

Amid all the detail, it was easy to miss Abrahamson's central message, which he brought up about a third of the way into his extensive opening remarks. In the context of a discussion about how it is useful to have different systems in place to perform the same functions, Abrahamson pointed to a slide and said: "For example, [we see here] a ground-based laser . . . going up and bouncing off a mirror in space and going forward to what we call a fighting mirror and then going down to destroy a missile in the boost phase. The advantage of that kind of a system is clearly that it strikes with the speed of light. On the other hand, you can't deliver as much energy. So therefore on the right-hand side you see something called an SBKKV, a space-based kinetic kill vehicle. Those are simple rockets in space."

Such rockets could be command-guided or they could rely on the infrared heat-seeking systems used in missile-defense and anti-satellite experiments in 1984 and 1985. In the September ASAT test, Abrahamson reminded the press, the Air Force "hit a satellite with something weighing a little less than fifty pounds. Our job now is to get down to less than ten pounds. We can use a rocket to get it up or use a railgun."

Asked following his formal presentation why it had been decided to decelerate or drop space-based lasers and railguns, Abrahamson said: "We didn't get the money that we needed, either in fiscal 1985 or 1986. And we had a choice. One choice is to try to take this broad range of technology and just slow it all down evenly. . . . The other one is to try to take the knowledge that we have developed in our experiments so far, in the research so far, in the architectural studies so far, and begin to make decisions. So we are making decisions. . . . So the way we're doing this is that we're not cutting out all of that research, we're just throttling way back to what I'm calling a backup technology."

The emphasis of the program, Abrahamson reiterated, would be on ground-based free-electron lasers and the "space-based kinetic kill vehicle." The space-based rocket "appears right now to be the simplest way to proceed," Abrahamson said, "and we see some very real cost reductions. . . . But in the future, we haven't cut out railgun activity, not at all. We can see that that one is more than just a backup technology, but we haven't proceeded with it at the rate at which I think there is real opportunity to proceed."

Abrahamson's announcement was not a major surprise in light of many developments. The Fletcher study team of defensive systems concluded two years ago that chemical lasers sould be emphasized only if there were no fiscal constraints on the SDI program. SID contractors were rumored to be uneager to make large R&D efforts in areas where they saw little or no chance for large equipment orders before the next century. An influential SDI panel on computer software recommended that a space-based ABM system consist of a very large number of coordinated but semiautonomous components. In keeping with that recommendation, Gerold Yonas, chief scientist of SDI, was known to be increasingly in favor of "pods"—space platforms carrying a small number of interceptors, which could be deployed by the hundreds or thousands.

Yonas denies that there has been a "sea change" in SDI and he says that some of what are normally thought of as Star Wars technologies will continue to be emphasized. He predicts, for example, that a neutral-beam weapon will be tested in space by the end of the decade, and he says that the SDI organization is trying to get added funds for nuclear-pumped x-ray-laser research, despite reports that recent test results have been of dubious validity. Yonas says the money is needed precisely in order to subject these results to closer scrutiny. . . .

The July 29, 1986, edition of the *Albuquerque Journal* contained an article claiming an announcement from the Air Force Weapons Laboratory at Kirtland Air Force Base that researchers have successfully synchronized the light from two independent lasers in what lab officials termed "a major step toward the development of a practical laser weapon." *Journal* science writer, Byron Spice, explains that "this phase-coupling" technique theoretically would allow an array of relatively small lasers and mirrors to inflict the same damage on a target as a very large, single laser.

The technique causes the intensity of laser radiation on a target to grow exponentially. That is, two lasers in phase with each other could deliver the brightness of four unphased lasers, three phased lasers the brightness of nine, four phased lasers the brightness of sixteen, etc.

The research isn't funded by the Pentagon's Strategic Defense Initiative Office but would have obvious applications in a missile defense system as well as in other weapon systems, said Lt. Ken McClellan, a lab spokesman.

The experiment was performed with two iodine lasers inside a building at Kirtland last August. The results only now have been released, however.

The work complements the lab's Phasar project, which concerns how to optically focus separate beams on a single target.

McClellan noted the phased array approach to laser weapons would enhance survivability. An array of seven phased lasers, equivalent to forty-nine unphased

MILITARY APPLICATIONS

lasers, would still have the brightness of thirty-six lasers if one of the lasers were destroyed or malfunctioned.

The lab has been studying the phase-coupling technique for about ten years. In 1982-83, researchers successfully coupled two carbon dioxide lasers.

The work with the iodine lasers was much more difficult and significant, McClellan said. The infrared light emitted by the iodine lasers has a wavelength ten times shorter than that of the carbon dioxide laser and thus requires ten times greater precision.

The iodine laser has some distinct advantages for weapons uses. Its light is produced by a chemical reaction, which means its source of energy is liquid fuel that can be stored compactly and for long periods of time, said Lt. Col. Thomas Walker, chief of the lab's quantum optics branch.

Also, the wavelength of the iodine laser light penetrates the atmosphere well, McClellan said.

Simply shining two or more lasers on the same spot doesn't achieve the same results as a phased array of lasers. The laser light not only must be the same wavelength, but the waves of light must be in step with each other—peaks of one wave matching the peaks of the other.

The weapons lab scientists achieved this lock-step by having the two lasers share some of their output with each other.

Though the coupling technique could be used for three or more lasers, a weapon employing the technology is still in the future. The laser brightness may increase exponentially as additional lasers are added, but so do the complexities, noted lab spokesman Rich Garcia.

Nevertheless, "this demonstration is a real breakthrough," Walker said. "It proves that we can phase-couple iodine lasers to obtain the high powers necessary for several Air Force Strategic Defense Initiative applications."

The following article on laser experiments set at White Sands was published in the *Los Alamos Monitor* on August 13, 1986.

> A major series of free electron laser experiments will be conducted at new facility at White Sands Missile Range, to evaluate laser designs being developed at Los Alamos and Lawrence Livermore National Laboratories, the Strategic Defense Initiative Office announced last week.
>
> The free electron laser is one of several weapons being considered as part of the proposed "Star Wars" defensive shield.
>
> Two different free electron laser systems are being studied at LANL and LLNL. LANL researchers are focusing on use of an radio frequency linear accelerator as the way to speed up a beam of electrons, while the work at Livermore concentrates on an induction linear accelerator, said a lab press release.

In the first phase of the experiments at White Sands, a ground-based free electron laser, using the LANL accelerator design, will be constructed. Later, higher power experiments will focus on the Livermore accelerator.

Submarine Tracking

In the November 14, 1985, issue of the *San Francisco Examiner*, an Associated Press article described advances in underwater laser technology.

> The Defense Department, in an experiment with significant implications for military strategy, has successfully transmitted messages via laser light from a high-flying airplane to a submarine crusing at "operational depths."
>
> The experiment, confirmed by Rear Adm. Thomas K. Mattingly and other Navy officials, was conducted more than a year ago off the coast of San Clemente, under the code name "SLCAIR (pronounced Slickair) 84."
>
> A small jet carrying an experimental green-light laser was able to establish contact and transmit messages "error-free" to a submerged submarine.
>
> Although precise details are classified, the airplane was flying at altitudes between 20,000 feet and 30,000 feet at the time of the transmissions, one source said. Another source said the term "operational depth" meant the submarine was more than 100 feet below the surface.
>
> The successful test has paved the way for additional research, and it has convinced some officials that a more advanced laser system can be constructed using satellites instead of airplances. Over the next two years, the Navy will take control of the research from the Defense Advanced Research Projects Agency, or DARPA.
>
> Although Navy officials caution that the service is still years away from building any operational system, the experiment offers one promising avenue for attacking a problem that has long dogged nuclear planners—how to communicate reliably with ballistic missile submarines without requiring the sub to rise near the surface and risk disclosing its position.
>
> Moreover, a laser communications system is viewed as having tremendous implications for tactical warfare, because it could allow surface ships to protect the whereabouts of U.S. attack submarines, while still directing them toward enemy submarines.
>
> The existence of the DARPA research program involving so-called blue-green lasers has long been public knowledge. The research

has been cited in the past by such lawmakers as Rep. Les Aspin, D-Wis., and Sen. Carl Levin, D-Mich., who see it as offering an alternative to the ELF (extreme low frequency) submarine communications system now being built in Wisconsin and upper Michigan.

Recently, however, Mattingly provided the first public acknowledgement the research had moved to the point of a successful transmission of data. A former astronaut who now directs space programs within the Space and Naval Warfare Command, Mattingly referred briefly to the test in an article he wrote for *Proceedings* magazine, published by the U.S. Naval Institute.

"Recently, the Navy and the DARPA demonstrated the use of an airborne laser to transmit data to a submarine at operational depths," Mattingly wrote in discussing research projects that could assist surface ships.

"A satellite-based global laser communication system promises the battle group commander the opportunity to communicate directly with his most effective anti-submarine warfare weapon, an attack submarine."

In an interview, Mattingly stressed the Navy and DARPA were still engaged in basic research "and not development of an operational system."

"But the experiment did demonstrate that you can, in fact, do it—have a submarine pick up messages under water and decode them," he continued. "They overcame the attenuation effects (of shooting the laser) through the atmosphere and water."

Air-to-Air Laser Gun

In a UPI article carried by the *Albuquerque Journal*, dated November 30, 1983, the successful destruction of a drone plane by an airborne laser was reported.

> An airborne laser beam destroyed a target drone and damaged two others in a test of its capabilities as a weapon, the Pentagon said Tuesday.
>
> The experiments conducted in September by the Air Force and the Navy marked the culmination of in-flight target testing of the Airborne Laser Laboratory, a Boeing 707 jetliner modified to carry a high-energy laser and designated an NKC-135.
>
> In the latest round of tests, a target drone was launched from the Navy's Pacific Missile Test Center at Point Mugu, Calif., and flown about twenty miles out to sea where it simulated a low-altitude attack against a Navy ship, a Pentagon statement said.

The NKVC-135 acquired the target with its on-board radars and 'engaged it with the laser,' the Pentagon said.

"The laser beam radiation caused target destruction when the laser burned through the skin of the drone and destroyed critical components, causing a flight control failure," it said.

Two other drones were damaged in similar tests.

There was no indication of the distance between the NKC-135 and the drone when the beam was fired at the target.

Army's Close Combat Laser Gun

It seems incredible that infantrymen might exchange their M-1's for laser guns, but this UPI article appeared in the *Albuquerque Journal*, dated December 18, 1983.

The Army is developing a controversial laser weapon that permanently blinds soldiers who look into it, even if they are looking through a tank's nightscope, it was reported Saturday.

The Washington Post, quoting defense industry sources, said the Army has asked for $14 million in 1984 to help develop the classified weapon, called the Close Combat Laser Assault Weapon. The Army has acknowledged that C-CLAW is under development but declined to discuss its potential, the *Post* said.

Pentagon spokesmen had no immediate comment.

The report said ethical questions about the weapon are being debated in Army circles, although for publication the Army insists there is no debate. The Soviets are believed to have already developed a similar weapon.

The portable C-CLAW could be mounted on tanks, personnel carriers, or even helicopters.

It is well known that laser-aiming devices on some police rifles have been experimented with and may be in use at this time, but beam intensities are too low to cause biological damage.

LASER-INDUCED FUSION

According to Harold Agnew, former Director of the Los Alamos National Laboratory, in the foreword of a pamphlet, "Introduction to Laser Fusion," published in 1975, scientists have been attempting to produce and control fusion reactions in "magnetic bottles. Another approach to fusion emerged shortly after the invention of the first lasers in 1960. This concept included the use of compression studies developed in the nuclear weapons program but would use intense laser

LASER-INDUCED FUSION

beams to drive spherical implosions of fusion fuel. The work was mainly theoretical until, in 1971, the Atomic Energy Commission greatly expanded the experimental activity, primarily at the national laboratories at Livermore, California, and at Los Alamos, New Mexico. The research has attained various degrees of success, but sustained thermonuclear reactions designed for electrical power generation is not anticipated in the near future."

The following information on the concept of laser-initiated fusion has been extracted from the pamphlet mentioned above.

Thermonuclear fusion is the process constantly occurring in the sun and also utilized in hydrogen weapon technology. *Controlled* thermonuclear fusion is the most technically demanding of all the concepts being investigated as sources of large amounts of power for peaceful uses. In terms of today's knowledge, fusion is an almost unlimited potential source of power, second only to the sun itself. Furthermore, it offers the possibility of eliminating, or at least substantially alleviating, many disadvantages of present and prospective methods for producing large amounts of power. More specifically, fusion power promises to be less polluting, safer, cheaper, and nearly inexhaustible. However, the harnessing of fusion power presents a tremendous scientific and engineering challenge. Even the most optimistic estimates do not include operation of the first commercial electrical power plant until near the end of this century.

Two approaches to fusion power are being pursued both in this country and abroad. The older of these schemes is based on magnetic confinement and heating of a plasma. The newer method is concerned with laser-initiated fusion. The two essential features that may eventually lead to laser-initiated fusion are design and construction of extremely powerful and efficient lasers and achievement of coupling, in a predetermined manner, between powerful laser beams and thermonuclear fuel targets, to produce appreciable thermonuclear burning of the target material.

A short-pulse laser process called "Q-switching" stimulates downward electron energy-level transitions that emit light in relatively short bursts. However, for laser fusion, the light pulses are too long (about one millionth of a second). Another process called "mode-locking" produces much shorter (about one billionth of a second) and more powerful light pulses that *are* of interest for laser fusion.

Recent technological advances toward shortening laser light bursts and making them more powerful have led to serious consideration of focusing such laser pulses onto pellets of thermonuclear fuel to initiate thermonuclear reactions that would release considerably more energy than that required for their initiation.

One way to prepare the DT fuel is to mix equal amounts of D and T in gaseous form and freeze the gas mixture into a tiny sphere or pellet about the size of buckshot. The DT ice pellet is then injected into an evacuated cavity.

When the pellit reaches the center of the cavity, its entire surface is illuminated as uniformly as possible by simultaneous laser beams from several directions.

Simultaneous illumination from various directions is achieved by starting with *one* laser beam source. The power of the laser beam from the source is amplified by another laser cavity, or preamplifier. The output beam from the preamplifier is then *split* into, say, eight beams, each of which is greatly reamplified in a power amplifier. All laser amplifiers work essentially on the principles discussed above; the incoming laser beams stimulate more photon emissions in each amplifier's lasing medium.

Note that light travels at a speed of 300 million meters per second. Therefore, a laser *pulse* (moving at the speed of light) that lasts for only a billionth of a second as it passes a given point is only about thirty centimeters (one foot) long!

The eight beams are directed symmetrically into the laser cavity by mirrors and lenses. The optics (the mirrors and lenses) are adjusted so that all eight beams (laser pulses) travel the same distance and thus strike the pellet at the same time to provide simultaneous symmetrical heating.

The main laser pulse must have the following characteristics to be efficiently absorbed by the pellet and produce fusion. It must be of the proper frequency, or color, and it must be short, lasting only about one billionth of a second. Even this short laser pulse must, furthermore, be *properly shaped*—of low power at first, increasing exponentially to more than a thousand million million watts of power—and it must have a total energy corresponding to that of about 227 grams (one-half pound) of high explosives, or roughly a stick of dynamite.

Absorption of the powerful main laser pulse may be aided by first illuminating the pellet with a small laser prepulse. The prepulse ionizes the pellet *surface*, forming a gas that expands relatively slowly to create an atmosphere around the pellet. This atmosphere helps the laser pulse to be absorbed by, rather than reflected from, the pellet surface.

Absorption of the main laser pulse then quickly heats the outer region of the pellet to form an ionized gas or plasma that expands outward (blows off) rapidly. The recoil impulse from the very rapid blowing off of the outer pellet layer compresses the pellet core in the same way that the impulse from a rocket's exhaust pushes the rocket forward, or a rifle-shot recoil pushes the rifle against one's shoulder.

Theory predicts that core compression should produce core temperatures of about 10 to 100 million degrees—about one to ten times hotter than the interior of the sun. High temperatures imply rapid particle movements in random directions. At such temperatures, the D and T are ionized, and the ion speeds are sufficient for fusion.

LASER-INDUCED FUSION

Theory predicts that the center of the pellet core will be compressed to superdensities, one to ten-thousand times the normal solid density, or about ten times the density of the center of the sun (about one hundred times as dense as lead). Such pellet-core densities are important because they greatly increase the likelihood that energetic Ds and Ts will collide with one another. Also, they enable still unfused D and T ions to recapture, or share, some of the energy of a fusion-product ^{4}He particle before the high-velocity ^{4}He particle can escape the core region. This is analogous to one fast-moving billiard ball striking and giving energy of motion to others. This energy sharing with unburned fuel gives rise to so-called "bootstrap" heating that further increases the reaction rate. Achievement of core compression is crucial to the laser fusion process.

Enough thermonuclear burning can occur to fuse an appreciable fraction of the D and T ions in the core very quickly, before outward motion of the energetic core material tears the pellet apart. In these circumstances, the thermonuclear reaction is said to be inertially confined.

Note again that the initiating laser pulse must be short and powerful; if the pellet were heated more slowly, the resulting plasma would have time to expand, losing density and heat before significant fusion could take place.

The thermonuclear burning phase is predicted to last only about ten picoseconds (one picosecond equals one millionth of one millionth of one second). Only part of the core DT needs to "burn" to create a net energy release.

In summary, when the powerful laser pulses shine onto the DT pellet, a miniature, short-lived sun, of even higher temperature and density than our Earth's sun, is created.

The enormously high temperatures produced by fusions in the pellet core cause the pellet material, now entirely an ionized gas or plasma, to explode. The energy released from the thermonuclear burning and microexplosion of such a tiny pellet should be about equal to that from 22.7 kilograms (50 pounds) of high explosives, giving a so-called gain factor of about 100 over the incident beam's original energy.

The energy yield from *each* D and T fusion is about 1000 times greater than the energy invested in the D and T atoms. But, because only a fraction of the heated DT fuel will be burned per pulse, the average *net* energy gain per pulse will, when laser efficiencies are also considered, be about ten; in other words, the fusion energy from the pellet will be about ten times greater than the energy needed to power the lasers.

Unlike that of a chemical explosion, the total mass associated with the fusion energy release is very small, but each particle has a *very* high energy (speed). Though the energy released would equal that from several dozen sticks of dynamite, the blast from a pellet microexplosion would only be like that from a large firecracker.

194 FUTURE DEVELOPMENTS

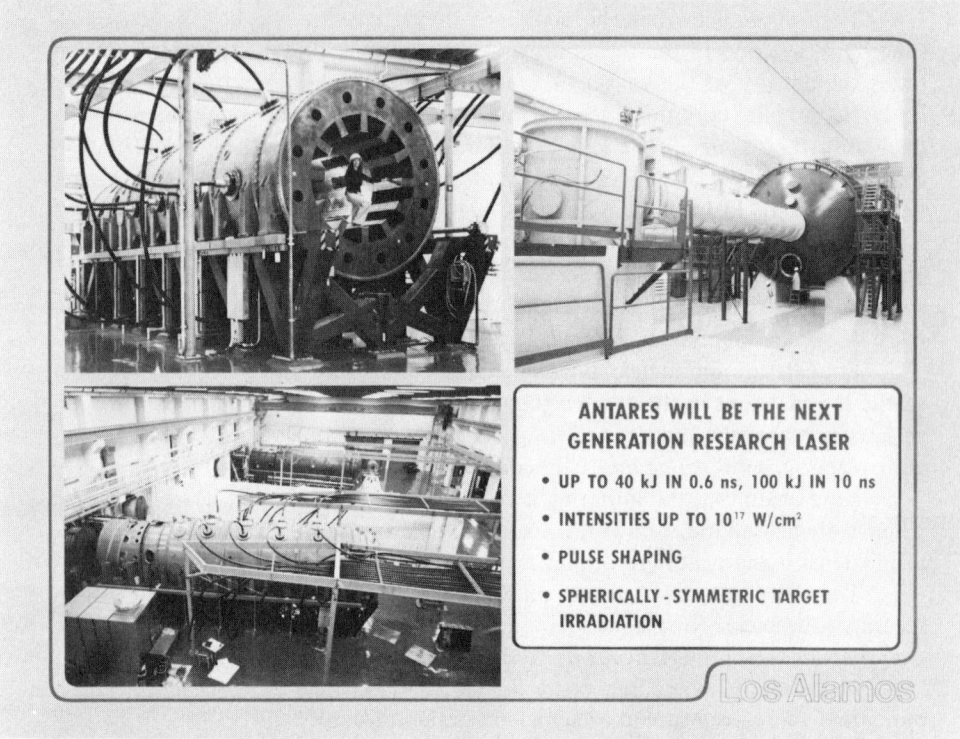

Figure 9.1 Views of the Antares laser facility, the world's largest CO_2 laser system. (Courtesy of Los Alamos National Laboratory, Los Alamos, New Mexico.)

The energy is released in various forms: as x rays (radiation from the plasma's free electrons), neutrons, ^{4}He particles, and "unburned" D and T ions. This energy must be captured and converted into electrical energy to recharge the lasers for the next pulse and to provide net electrical energy output for consumers.

Since research on laser fusion began at an accelerated pace in the early 1970s, evidence of obtaining the fusion reaction has been claimed by several sources. However, the intensity of the 14 meV electrons resulting from the impingement of laser energy focused on the small targets containing DT has been insufficient to encourage scaling up of the huge lasers used in the experiments with the CO_2 laser system at Los Alamos (see Fig. 9.1), and with the solid-state Nd:glass laser system at the Lawrence Livermore Laboratory. Recent experiments have used shorter wavelength laser energy, either by trying new processes to generate shorter wavelengths (in ultraviolet region) such as the excimer gas lasers that could utilize facilities at Los Alamos, or by using doubling techniques to obtain harmonic vibrations (shorter wavelengths) of the solid-state system at Livermore. Research is continuing at these facilities of the University of California operating under contract to the Department of Energy.

NUCLEAR LASER

A device called a "graser" (acronym for *g*amma *r*ay *a*mplification of *s*timulated *e*mission of *r*adiation) is being investigated by scientists at Los Alamos National Laboratory. The following article by Jack Challem appeared in the May 9, 1986 edition of the *Los Alamos Newsbulletin.*

    A team of Los Alamos scientists has successfully completed the first in a crucial series of experiments that may eventually lead to the world's first nuclear laser.

    The device, called a graser, is the focus of a speculative, high-risk project that would produce a beam of intense gamma rays much the way conventional lasers produce intense beams of light.

    Despite the graser's many "ifs," researchers say it could be a boon to biological research, medical therapy, and worldwide communications.

    "At some point, someone is going to make a gamma-ray laser," said Peggy Dyer of Subatomic Research and Applications (P-3). "It's just a question of whether it will take three years or thirty years."

    Like lasers used in compact disk players, grasers would emit energy in a concentrated beam. But unlike these and other lasers, a graser would make direct use of nuclear energy and produce a thin beam of gamma rays, which have a very short wavelength. These features would make it extremely powerful.

    But the future of gamma-ray lasers depends on the successful completion of a series of basic "proof-of-principle" experiments in largely uncharted scientific areas. These experiments would lay the technical foundation for the futuristic instrument.

    Dyer and P-3's George Baldwin have taken one of the necessary first steps.

    The gamma-ray laser concept they are following uses a material called an isomer, which would be implanted in a crystal. An isomer is an energetically excited state of an atom's nucleus.

    That crystal would keep the isomeric nuclei in a defined pattern to ensure the graser's performance. It would absorb nuclear recoil to keep the nuclei in resonance, as well as to align the gamma-ray beam.

    Until now, though, no one has been able to separate an isomer from an isotope of the same element. An isomer has the same mass, or weight, as an element's isotope but a slightly different structure.

The researchers' proof-of-principle experiments used mercury 197. With beams from conventional lasers, Dyer and Baldwin "excited' mercury atoms containing the isomeric nuclei. When the atoms vibrated from the effect of the laser beam, the isomers broke free.

"The experiment was a handle to grab on to," said Dyer. "A person couldn't even think of building a gamma-ray laser without being able to get the isomer."

The researchers' next step is to learn how to incorporate, or dope, an isomer into a crystal. Such crystals are the heart of most gamma-ray laser concepts, and the crystals' composition and quality will largely determine graser performance.

The final and most difficult obstacle for Dyer and Baldwin will be developing a way to feed energy into the isomer without destroying the crystal's delicate structure. Only with that accomplished will it be possible to achieve a gamma-ray laser.

Baldwin and Dyer say the gamma-ray laser might be used to create three-dimensional holographic images of living body cells, in essence becoming a sophisticated microscope.

Like many of today's lasers, a graser could also perform delicate surgery in place of scalpels and other surgical tools.

Finally, properties unique to the graser would make it suitable for rapidly transmitting large numbers of messages around the world by satellite.

Those potential applications please Baldwin, who, after researching grasers for twenty years, is described by colleagues as the father of the gamma-ray laser in the United States.

"In 1958," he said, "no one could have predicted the widespread use of lasers in surgery, fusion energy, computer printers, or even home entertainment. A new invention's most significant applications are rarely foreseen."

Added Dyer, "Whenever you do something hard, like try to develop a gamma-ray laser, you run into interesting physics. If it doesn't work, it's not as though you haven't done anything worthwhile. We always learn new things."

10
Summary

Laser technology is being employed in essentially all scientific and engineering endeavors. Engineers and technicians are being called on to determine what processes or procedures can produce the most efficient and the most cost effective products and services, and whether laser beams can be employed to help. Before such recommendations can be made to management, knowledge of how laser beams can be applied to various processes or procedures must be obtained in order to select the appropriate laser system. Such beam characteristics as wavelength, mode of propogation, beam intensity, and target material interaction must be understood in order to determine the beam's possible usefulness. Even after deciding on procurement of a particular laser system, beam measurements must be constantly monitored to maintain beam integrity, so that constancy of results can be ensured. Chapters 2, 3, and 4 contain information vital to be understood.

Before a laser is placed into operation, the emitted beam must be evaluated for hazard control. Following the required procedures described in Chapter 5 from the ANSI Z136.1 *Safe Use of Lasers* standard can prevent possible accidents not only from the beam's hazard but from other hazards associated with the laser operation also, such as electrical, chemical, compressed gases, toxic materials, etc. It is particularly important to follow control procedures required for lasers employed in light shows and obtain approval from state and local agencies. A complete list of these agencies is listed in Chapter 8.

SUMMARY

The transfer of energy by the stimulated emission of radiation amplified in a laser cavity to produce a coherent beam of photons can be applied to a plethora of reactions. This energy can be focused to a very small spot for high intensities required for materials processing such as cutting, welding, drilling, etc. The energy can be transported along optical fibers to provide efficient communication channels thousands of times more effective than electronic systems. By processing a beam with optical systems, digital codes can be "read" by holographic techniques, and three-dimensional evaluation of objects by non-destructive methods can be employed. Although an extensive variety of lasers are currently available commercially, new lasing methods involving numerous materials are constantly evolving to produce advanced laser technology. It behooves the practitioner to adapt the most efficient and most cost effective technique involving lasers to the process at hand. Hopefully, the applications described in Chapters 6, 7, and 9 will prove inspirational. Keeping current with this dynamic and indispensable industrial and research tool requires subscription to periodicals published by specialists in various fields, especially by the laser electro-optical associations. Annual conferences are provided at domestic and foreign locations to offer an exchange of information on various specialties in laser technology. Several of these organizations are listed below:

Laser Institute of America, 5151 Monroe Street, Toledo, OH 43623, (419) 882-8706

Optical Society of America, 1816 Jefferson Place, NW, Washington, DC 20036, (202) 223-8130

SPIE-Society of Photo-Optical Instrumentation Engineers, P.O. Box 10, Bellingham WA 98227-0010, (206) 676-3290

By directing the energy carried by a controlled stream of photons in a coherent array, there appears to be an inexhaustible number of varied applications that can convert that energy to useful work. The only limitations seem to be the users imagination and inventiveness.

Index

A

Air Force Weapons Laboratory at Kirtland Air Force Base, 186
Applications of lasers in science and engineering, 97-140
  chemistry, 118-124
    laser induced spectroscopy, 118-119
    special isotope separation, 119
    ultrafast spectroscopy, 123
  fiber optics communications, 109-115
  halography, 115-118
  materials processing, 98-102
    cutting and drilling, 100
    heat treatment, 100
    other, 109
    welding, 100
  semiconductor processing, 103-109
    laser writing, 107
    photolithography, 107

Astronomer's measurements with lasers, 129

B

Beam characteristics, 21
Beam cross-sections, or modes, 38-42
Beam divergence, 42-43
Beam focusing, 43-47
  diffraction-limited spot size, 45-47
Beer's law, 61
Biological damage mechanisms, 85
Biological damage threshold values, 87

C

Chemical lasers, 13
  hydrogen fluoride (HF), 13
  iodine, 186

Continuous beam (CW) properties, 34-35

E

Einstein's theory of stimulated emission, 5-6
Electromagnetic spectrum, 21-32
  frequency, 22-23
  interference of light waves, 32-33
  phase, 26, 27
  polarization, 27-32
  wavefronts, 26, 27
  wavelength, 22-23, 33
  wave theory, 25
Energy density, 84
Engineering controls for laser hazards, 88
  interlocks, 90
  lights, 90
  protective eyewear, 90
    Fred Reed Optical Company, 90-91
  signs, 89
Excimer lasers, 17
  krypton fluoride (KrF), 17
  mercuric halide, 20

F

FDA division of compliance, 174
Free electron laser, 18, 158-163
Future developments of laser technology, 177-196
  automobile navigation, 177-179
  gamma-ray laser, 195-196
  laser-induced fusion, 190-195
  military applications, 179-190
    air-to-air laser gun, 189
    laser weapon in space, 179
    star wars technology, 180

[Future developments of laser technology]
  strategic defense initiative, 183-188
  submarine tracking, 188-189

G

Gamma-ray laser, 195-196
Gas lasers, 11-13
  argon, 12, 108, 127, 150
  carbon dioxide (CO_2), 11, 12, 148
  excimers, 103-109, 154
  helium-neon (HeNe), 11
  krypton, 12
  oxygen, 20

H

History of light, 2
Human eye transmission curve, 84
  wavelengths absorbed by eye components, 86

I

Image converter tube, 79, 80
Infrared telescope, 79

J

Joule/watt relationship, 86

L

Laser absorption in human tissue (chart), 142

INDEX

Laser alignment system, 131
Laser as an art form, 165-176
  laser light shows, 165-166
    FAA regional offices, 170
    individual responsibilities, 174
    light show government requirements, 167-174
    open air light shows and FAA regulations, 169-170
    state and local requirements, 170-174
    state radiation control offices, 171-174
Laser cavity, 9
Laser conference sponsors, 198
Laser detector for search and rescue, 132
Laser disc recorders and players, 138-139
Laser fingerprint identification, 135-136
Lasers, free electron, 187
Laser gyro, 133-134
Laser safety eyewear selection chart, 93
Laser safety indoctrination, 91-93
Laser safety officer, 92-95
Laser safety program, 93-96
Laser sound recovery from antique audio cylinders, 136-138
Laser tunnelling, 131
Laser video discs, 139-140
Laser wind-sensor, 134
Lasers in computers and printers, 124-128
  diode laser printer, 127
  photoplotters, 127
  reprographics, 127
Lasers in construction, 124
Lasers in shock wave diagnostics, 133

Liquid lasers, 15
  organic dyes, 15
Lawrence Livermore National Laboratory, 9, 184, 187, 191, 194
Los Alamos National Laboratory, 12, 14, 18, 118, 119, 122, 131, 132, 133, 156, 180, 181, 187, 190, 194, 195

M

Maiman, Theodore, laser inventor, 8
Market predictions, 1-3
Measurement of laser beam characteristics, 49-82
Measuring Doppler effect with lasers, 130
Medical applications of lasers, 141-163
  arthritic syndromes, 158
  cell sorting, 156-158
  dermatology, 155-156
  laser surgery, 141, 143-150
    brain tumors, 1456
    cauterizing blood vessels, 148
    coronary angioplasty, 145-148
    endoscopy, 148
    general surgery, 148
    kidney stones, 148-149
    podiatric treatments, 145
    rectal conditions, 145
    tissue welding, 143-144
  ophthalmology, 150-155
    cataract, 153
    diabetic retinopathy, 150
    detached retinas, 156
    glaucoma, 152
    mocular degeneration, 150, 152
    radial keratotomy, 156
Metal vapor lasers, 19

Metric prefixes, 24
Micro-wire stripping with lasers, 128

N

Noise, 66-67
Nuclear laser, 195

O

Ocular focus wavelengths, 84

P

Phase-coupling, 186
Photographic instrumentation, 76-82
  ASA film speed, 77
  Dynafax camera, 78
  multi-framing camera, 78
  oscilloscope photography, 76
  polaroid film, 77
  prefogging in film, 77
  streak camera, 78, 79
Photolithography, 107
Power and energy measurements, 62-76
  photodetector, 62-67, 72
    quantum detectors, 63
    thermal detectors, 64, 69-71
  photometric units, 68-69
  power meter, 70
  radiometric units, 68-69
Power density, 84
Pulsed laser beam traits, 35-36
  normal pulsed mode, 36
  Q-switching, 37
  repetitive pulses, 38
  rotating prism, 36-37
  ultrafast pulsing, 38

Q

Quantum efficiency, 66

R

Response time, 64
Retinal hazard wavelengths, 84

S

Safe use of lasers standard, ANSI Z 136.1, 88-91
Safety in the laser environment, 83
Sandia National Laboratory, 126
Semiconductor lasers, 9-11
Semiconductor processing, 103-109
Sniperscope, or snooperscope, 80
Solid-state lasers, 7
  Alexandrite, 119, 121
  diode, 123
  injection, 125, 156
  ruby laser, 7, 8, 9
  semiconductor, 103-109, 126
  tunable, 9
  YAG laser, 7, 9, 118, 148, 153
Speckle, 104
Stanford University, 18

T

Tunable lasers, 9, 10

W

Wavelength chart showing laser locations, 85

INDEX

Wavelength measuring devices, 49-62
  grating spectrometer, 51
  monochromator, 54-61
  prism spectrometer, 50
  spectrograph, 49
  spectrometer, 49
  spectrophotometer, 62

[Wavelength measuring devices]
  spectroscope, 49

X

X-ray laser, 19

About the Author

D. C. WINBURN is a consultant for several firms involved in laser technology and laser manufacturing. In 1982 he retired from the Los Alamos National Laboratory in New Mexico after 35 years in a variety of technical and administrative positions, including technical administrative assistant to the Laser Research and Technology Division (1972-82). A recognized authority on laser safety, he is the author of numerous articles and one book, *Practical Laser Safety* (Marcel Dekker, Inc.). From 1975 to 1980 he served as Secretary of the Laser Institute of America, and since 1973 he has been a member of the American National Safety Institute's ANSI-Z136.1 "Safe Use of Lasers" Committee. Mr. Winburn has lectured at the New York Academy of Sciences, American Industrial Hygiene Association, American Society of Safety Engineers, Japanese Conference on Lasers (1976), Australian Naval Experimental Laboratory, and at numerous colleges and universities. He served on the faculty of the University of Cincinnati Medical School's Short Course on Laser Safety, and he was an advisory committe member for laser technology at Idaho State University at Pocatello, University of New Mexico at Los Alamos, and Technical Vocational Institute in Albuquerque. He is a member of the Optical Society of America, Laser Institute of America, and Society of Photo-Optical Instrumentation Engineers. Mr. Winburn received the B.S. degree (1943) in metallurgical engineering from the South Dakota School of Mines and Technology in Rapid City.